Selected Titles in This Series

(Continued in the back of this publication)

Cocycles of CCR Flows

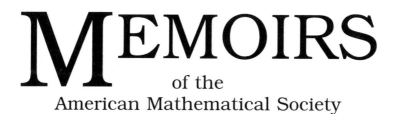

MEMOIRS
of the
American Mathematical Society

Number 709

Cocycles of CCR Flows

B. V. Rajarama Bhat

January 2001 • Volume 149 • Number 709 (end of volume) • ISSN 0065-9266

American Mathematical Society
Providence, Rhode Island

1991 *Mathematics Subject Classification.*
Primary 46L55, 81S25.

Library of Congress Cataloging-in-Publication Data

Bhat, B. V. Rajarama, 1966–
    Cocycles of CCR flows / B. V. Rajarama Bhat.
        p. cm. — (Memoirs of the American Mathematical Society, ISSN 0065-9266 ; no. 709)
    "January 2001, Volume 149, Number 709 (end of volume)."
    Includes bibliographical references.
    ISBN 0-8218-2632-8 (alk. paper)
        1. Stochastic processes. 2. Quantum theory. 3. Cocycles. 4. Dilation theory (Operator theory) I. Title. II. Series.
QA3 .A57 no. 709
[QC174.17.S76]
510 s—dc21
[530.12′01′51923]   00-046915

Memoirs of the American Mathematical Society

This journal is devoted entirely to research in pure and applied mathematics.

Subscription information. The 2001 subscription begins with volume 149 and consists of six mailings, each containing one or more numbers. Subscription prices for 2001 are $494 list, $395 institutional member. A late charge of 10% of the subscription price will be imposed on orders received from nonmembers after January 1 of the subscription year. Subscribers outside the United States and India must pay a postage surcharge of $31; subscribers in India must pay a postage surcharge of $43. Expedited delivery to destinations in North America $35; elsewhere $130. Each number may be ordered separately; *please specify number* when ordering an individual number. For prices and titles of recently released numbers, see the New Publications sections of the *Notices of the American Mathematical Society*.

Back number information. For back issues see the *AMS Catalog of Publications*.

Subscriptions and orders should be addressed to the American Mathematical Society, P. O. Box 845904, Boston, MA 02284-5904. *All orders must be accompanied by payment.* Other correspondence should be addressed to Box 6248, Providence, RI 02940-6248.

Memoirs of the American Mathematical Society is published bimonthly (each volume consisting usually of more than one number) by the American Mathematical Society at 201 Charles Street, Providence, RI 02904-2294. Periodicals postage paid at Providence, RI. Postmaster: Send address changes to Memoirs, American Mathematical Society, P. O. Box 6248, Providence, RI 02940-6248.

Contents

Abstract

We study the partially ordered set of quantum dynamical semigroups dominated by a given semigroup on the algebra of all bounded operators on a Hilbert space. For semigroups of $*$-endomorphisms this set can be described through cocycles. This helps us to prove a factorization theorem for dilations and to show that minimal dilations of quantum dynamical semigroups with bounded generators can be got through Hudson-Parthasarathy cocycles.

2000 Mathematics Subject Classification : 46L57, 81S25

Key words and Phrases : minimal dilations, completely positive maps, CCR flows, E_0-semigroups, cocycles.

The author is supported by the Indian National Science Academy under Young Scientist Project.

Received by the editor February 5, 1998.

This work is dedicated to
Professor K. R. Parthasarathy

1 Introduction

This article builds a bridge between two subjects which have been develop-
ing more or less independently since eighties. The first one is the theory
of quantum stochastic processes whose foundation stones were laid by the
introduction of abstract notion of non-commutative processes as families
of *-homomorphisms by Accardi, Friegerio, and Lewis [AFL]. The subject
took a more concrete shape with the discovery of quantum Ito formula by
Hudson and Parthasarathy [HP]. Since then this has been a very fertile area
for research. This field has interactions with Probability, Operator theory
and Physics and the literature is really vast. We give two books Meyer
[Me], Parthasarathy [Pa] and one lecture notes in French [Bi] as basic references for this subject. We will be using the notation of [Pa]. The second
topic is the study of E_0-semigroups or semigroups *-endomorphisms of the
algebra $\mathcal{B}(\mathcal{H})$ of all bounded operators on a Hilbert space. (More general
von Neumann algebras also have been considered). The research in this
topic was initiated by R. T. Powers [Po1,2] in late eighties. Here [Ar1-7],
[Po1-4], [PP], [PR] are some basic references. More references can be found
in these papers. It is surprising that though the structure of semigroups
of automorphisms of $\mathcal{B}(\mathcal{H})$ was studied right in the begining of quantum
theory (Stone's theorem, spectral theorem and Hahn-Hellinger theorem explain almost everything), not much attention had been given into the study
of semigroups of endomorphisms. What is even more surprising is that
even after a decade, many beautiful papers of Arveson, Powers and others
not withstanding, we know very little about the structure of E_0-semigroups
other than CCR flows. CCR (Canonical Commutation Relation) flows are

paradigm examples of E_0-semigroups and one obtains them by considering second quantized shift operators on the Boson Fock space $\Gamma(L_2(I\!R_+, \mathcal{K}))$ of $\mathcal{K}$ valued square-integrable functions ($\mathcal{K}$ being a non-zero Hilbert space).

Appearence of E_0-semigroups in quantum stochastic calculus was noticed already in [Br]. A definite link between two subjects was found in [Bh2] where the theory of weak Markov flows developed in [BP1-2] was used to show that semigroups of completely positive maps get dilated to semigroups of $*$-endomorphisms in a canonical fashion. Semigroups of completely positive maps are known as quantum dynamical semigroups and there is lot of literature on them (see [AL], [Ch], [Da1-4], [EL]). By now Arveson has studied dilation in the context of E_0-semigroups. For example, he has shown that quantum dynamical semigroups (on $\mathcal{B}(\mathcal{H})$) with bounded generators gets dilated to E_0-semigroups cocycle conjugate to CCR flows [Ar5]. (Powers [Po5] has a different proof of this result when $\mathcal{H}$ is finite dimensional). However, as one knows the structure of these generators explicitly (thanks to Lindblad, Evans and Christensen ([GKS], [Li], [EC])) one would like to write down the cocycles also explicitly. Now we show that this *is* possible using the theory of quantum stochastic differential equations of Hudson and Parthasarathy([HP], [Pa], [Me]). This links the two subjects even more closely as HP theory is really the mainstay of quantum stochastic calculus.

Actually Hudson-Parthasarathy theory gives many candidates for dilation of a given quantum dynamical semigroup. The problem is to identify the right one (the minimal one). Checking minimality analytically seems to be rather hard (see [BF] for some special cases). Here we have a completely algebraic method. Consequently we know the structure of CCR flows even better. For example, it is an immediate corollary of these results that any non-trivial dilation which is cocycle conjugate to CCR flow of index 1 (that is, dim $(\mathcal{K}) = 1$) is automatically minimal (Theorem 8.7).

In the context of quantum stochastic calculus we answer some fundamental questions. For example to consider dilations of a given quantum dynamical semigroup why should one be tensoring the initial space with a Fock space. Here we show that it is more or less automatic that such a factorization exists (Theorem 8.9). Moreover, at least when the semigroup has bounded generator the unique minimal dilation associated with the semigroup always satisfies a quantum stochastic differential equation.

Now we go on to describe our methods and at the same time explain the arrangement of sections. In Section 2 we make precise our definition of conjugacy and cocycle conjugacy. Theorem 2.5 is very important for it shows that two notions coincide for primary dilations. In Section 3 we give the basic set up and notation with which we work throughout. One important notion introduced is the notion of induced semigroup. (Perhaps, this was there in the background in some of Arveson's work [Ar3] as well). Roughly speaking by knowing the induced semigroup one knows as to how far the given dilation is from being minimal. Theorem 3.7 shows that in some cases dilations are automatically minimal.

In Section 4 the partial order of domination for quantum dynamical semigroups has been introduced. One interesting fact here is Proposition 4.2, where positivity implies complete positivity, which is a rare thing to happen in non-commutative contexts. The main result of this section is Theorem 4.3, which says that in case of E_0-semigroups dominated semigroups can be described through positive, contractive, local cocycles. This is a key result. The collection of such cocycles forms a cocycle conjugacy invariant for the E_0-semigroup. Dominated semigroups can be thought of as abstract Feynman-Kac perturbations of the E_0-semigroup. In the next section we study how dominated semigroups behave under compression or dilation. The important fact which emerges is that we can detect minimality of the dilation by looking at the action of compression map on the partially ordered set of dominated semigroups.

The concept of units is very basic for the classification theory of E_0-semigroups. In Section 6 we give several descriptions of units. We also study as to what happens to units under dilation. It was already observed by Arveson [Ar4] that units of quantum dynamical semigroups can be lifted to their dilations. Here we give an explicit formula (6.2) for such a lifting. It is shown that units are in 1-1 correspondence with certain extensions of quantum dynamical semigroups. This helps us to answer a question raised by Davies long time back in [Da4].

In Section 7 we describe quantum dynamical semigroups dominated by CCR flows by computing all positive, contractive, local cocycles. It is a pleasant surprise that they have a very neat form. The Heisenberg motion group is the group of unitary local cocycles of CCR flows and this group acts on the set of positive, contractive local cocycles. An E_0-semigroup is said to be amenable if it has sufficient number of unitary local cocycles (Definition

8.2). It is shown that amenable dilations of elementary semigroups factorize uniquely in certain fashion (Theorem 8.6). An immediate corollary of this is that Powers' standard form is unique for amenable E_0-semigroups. It is also shown that type I dilations factorize uniquely as minimal dilation tensored with some other E_0-semigroup in standard form (Theorem 8.9). Perhaps these factorization theorems are the most important results of this paper.

Section 9 is an application of almost all the results of previous sections and it exhibits the bridge between E_0-semigroup theory and quantum stochastic calculus alluded in the beginning. We show that dilations of unital quantum dynamical semigroups with bounded generators can be realized through Hudson-Parthasarathy cocycles. In fact, we are able to compute 'deficiency index' which measures how far the given dilation is from being minimal. As a consequence an explicit description of the set of generators of quantum dynamical semigroups dominated by a fixed semigroup with bounded generator can be obtained. Heisenberg motion group (of appropriate dimension) has a natural action on it. A technical result of Parthasarathy and Sunder [PSu] on exponentials of indicators on Fock spaces is a simple corollary.

Finally there are two appendices. Appendix A tells us that often quantum dynamical semigroups and cocycles of E_0-semigroups are continuous if they are measurable. In Appendix B a decomposition of non-minimal dilations in discrete time has been outlined.

2 Compressions and dilations

Let $\mathcal{H}$ be a complex separable Hilbert space and let $\mathcal{B}(\mathcal{H})$ be the von Neumann algebra of all bounded operators on $\mathcal{H}$. By a quantum dynamical semigroup on $\mathcal{B}(\mathcal{H})$ we mean a one parameter semigroup $\alpha = \{\alpha_t : t \geq 0\}$ of (linear) completely positive maps mapping $\mathcal{B}(\mathcal{H})$ to itself. Throughout this article we assume that our semigroups are such that for every t, α_t is contractive and normal and for every $X \in \mathcal{B}(\mathcal{H})$ the map $t \mapsto \alpha_t(X)$ is continuous in the weak operator topology. There is a distinguished class of quantum dynamical semigroups consisting of endomorphisms. If α_t is a $*$-endomorphism of $\mathcal{B}(\mathcal{H})$ for every t, then α is called on e_0-semigroup. If further every α_t is unital then it is called an E_0-semigroup. If α_t is of the form $\alpha_t(\cdot) = A_t(\cdot)A_t^*$, for a strongly continuous semigroup of contractions $\{A_t : t \geq 0\}$ then α is called *elementary*. Note that if an E_0-semigroup is elementary then it is a semigroup of automorphisms.

Just the same way as one talks of isometric dilations of contractions one can talk of E_0-semigroup dilations of quantum dynamical semigroups. Suppose $\tau = \{\tau_t : t \geq 0\}$ is a quantum dynamical semigroup of $\mathcal{B}(\mathcal{H}_0)$. If $\mathcal{H}$ is a Hilbert space containing $\mathcal{H}_0$ as a closed subspace and if $\theta = \{\theta_t : t \geq 0\}$ is an e_0-semigroup of $\mathcal{B}(\mathcal{H})$ such that

$$\tau_t(X) = P\theta_t(X)P, \quad t \geq 0, \quad X \in \mathcal{B}(\mathcal{H}_0) = P\mathcal{B}(\mathcal{H})P \subseteq \mathcal{B}(\mathcal{H}) \qquad (2.1)$$

where P is the orthogonal projection of $\mathcal{H}$ onto $\mathcal{H}_0$, then θ is called a *dilation* of τ and τ is called a *compression* of θ. Note that even if θ is just a quantum dynamical semigroup (not necessarily an e_0-semigroup) given any projection

6

P in $\mathcal{B}(\mathcal{H})$ it is possible to define a family $\tau = \{\tau_t : t \geq 0\}$ of completely positive maps of $\mathcal{B}(\mathcal{H}_0)$ with $\mathcal{H}_0 = $ range P, taking (2.1) as definition. Such a τ would again be called as a compression of θ by P, though in general it may not be a quantum dynamical semigroup. But when it comes to dilations we are interested in the special case mentioned above. So in the sequel dilation would always mean e_0-semigroup dilation of a quantum dynamical semigroup unless we specifically mention otherwise.

The dilation θ is said to be *minimal* if the subspace generated by its action on $\mathcal{H}_0$ is $\mathcal{H}$, that is, if $\hat{\mathcal{H}} :=$

$$\overline{\text{span}}\{\theta_{r_1}(X_1)\cdots\theta_{r_n}(X_n)u : r_i \geq 0, X_i \in \mathcal{B}(\mathcal{H}_0), u \in \mathcal{H}_0, 1 \leq i \leq n, n \geq 0\}$$

is whole of $\mathcal{H}$. In the context of non-commutative Markov processes dilation means nothing but the construction of process starting with transition functions. The concept being so basic it has been looked at by many. We refer for example to [Be], [Em], [Ku1-2], [Sa], [Vi]. Making use of [Bh1], [BP1-2], it was shown in ([Bh2-3]) that *every* contractive quantum dynamical semigroup of $\mathcal{B}(\mathcal{H}_0)$ admits a unique (up to unitary equivalence) minimal dilation. Moreover the minimal dilation is an E_0-semigroup if the original semigroup is unital. Though in general minimal dilation need not be acting on a separable Hilbert space, once we assume that (as we do in this article) τ_t is normal for every t, and that the map $t \mapsto \tau_t(X)$ is weak operator continuous for every X, this is the case. (See Appendix A, Proposition A.2)

Let θ be an E_0-semigroup of $\mathcal{B}(\mathcal{H})$. A projection P in $\mathcal{B}(\mathcal{H})$ is said to be *increasing* for θ if $\theta_t(P) \geq P$ for all t. If P is increasing using the endomorphism property of θ_t it is easily verified that the compression τ of θ by

P (defined by (2.1)) is a unital quantum dynamical semigroup, so that θ
is a dilation of τ. The projection P and dilation θ are called *primary* if
$\theta_t(P) \uparrow 1$ as t increases ∞. If θ is a minimal dilation of τ then clearly P is
primary. However the converse is not true, and this can be seen by simple
examples. We give a discrete semigroup example here. Though we mostly
deal with one parameter semigroups the theory would be applicable for the
discrete case with very little modification.

Example 2.1: Let $\mathcal{H}_0$ be a Hilbert space with $\dim(\mathcal{H}_0) \geq 2$. Fix a unit
vector e in $\mathcal{H}_0$ and let $\mathcal{H}$ be the infinite tensor product $\mathcal{H}_0 \otimes \mathcal{H}_0 \otimes \cdots$, taken
with respect to the stabilizing sequence $\{x_n\}$ with $x_n = e$ for all n. Now $\mathcal{H}_0$
can be thought of as a subspace of $\mathcal{H}$ by identifying a vector $x \in \mathcal{H}_0$ with
$x \otimes e \otimes e \otimes \cdots$ in $\mathcal{H}$. Consider the normal $*$-endomorphism θ of $\mathcal{B}(\mathcal{H})$ defined
by

$$\theta(X_1 \otimes X_2 \otimes \cdots X_n \otimes I \otimes I \otimes \cdots) = I \otimes X_1 \otimes X_2 \otimes \cdots \otimes X_n \otimes I \otimes I \cdots$$

for $X_i \in \mathcal{B}(\mathcal{H}_0)$, and extended by normality, linearity to whole of $\mathcal{B}(\mathcal{H})$. The
orthogonal projection P of $\mathcal{H}$ onto $\mathcal{H}_0$ is given by $P = I \otimes P_0 \otimes P_0 \otimes \cdots$,
where $P_0 = |e\rangle\langle e|$ (the projection onto $\mathbb{C}e$). It is clear that P is increasing
for the semigroup $\{\theta^n : n \geq 0\}$. The compression gives us the semigroup
$\{\tau^n : n \geq 0\}$ of $\mathcal{B}(\mathcal{H}_0)$, where

$$\tau(X) = \langle e, Xe \rangle I, \quad X \in \mathcal{B}(\mathcal{H}_0).$$

It is not difficult to see that $\{\theta^n : n \geq 0\}$ is the minimal dilation of $\{\tau^n : n \geq 0\}$. $\{\theta^{2n} : n \geq 0\}$ is also a dilation of $\{\tau^n : n \geq 0\}$, however it is not minimal.
Note that the projection P is primary for the semigroup $\{\theta^{2n} : n \geq 0\}$ as
well as for the semigroup $\{\theta^n : n \geq 0\}$.

We remark that in Example 2.1 if $\mathcal{H}_0$ is infinite dimensional $\{\theta^n : n \geq 0\}, \{\theta^{2n} : n \geq 0\}$ are clearly conjugate (they are identical) as E_0-semigroups. However they are distinct as dilations of $\{\tau^n : n \geq 0\}$ as one of them is minimal and the other one is not. Similar situation prevails in the continuous case as well. So one needs to be careful while talking about conjugacy or cocycle conjugacy of dilations. We make the necessary definitions here and then recall classification scheme of Powers and Arveson for E_0-semigroups.

Suppose that θ, θ' are E_0-semigroups acting on $\mathcal{B}(\mathcal{H}), \mathcal{B}(\mathcal{H}')$, and are dilations of unital quantum dynamical semigroups τ, τ' acting on $\mathcal{B}(\mathcal{H}_0), \mathcal{B}(\mathcal{H}_0')$ respectively.

Definition 2.2: E_0-semigroups θ, θ' are said to be *conjugate* if there exists a unitary $M : \mathcal{H} \to \mathcal{H}'$ such that

$$\theta_t'(Z) = M\theta_t(M^* Z M)M^* \quad \forall Z \in \mathcal{B}(\mathcal{H}'), \ t \geq 0.$$

If further M can be chosen such that $\mathcal{H}_0' = M(\mathcal{H}_0)$ and

$$\tau_t'(X) = M\tau_t(M^* X M)M^* \quad \forall X \in \mathcal{B}(\mathcal{H}_0') \subseteq \mathcal{B}(\mathcal{H}'), t \geq 0,$$

then θ, θ' are said to be *conjugate as dilations* of τ, τ'.

Definition 2.3: A strongly continuous family of operators $G = \{G_t \in \mathcal{B}(\mathcal{H}) : t \geq 0\}$ is said to be a *(left) cocycle* for an E_0-semigroup θ of $\mathcal{B}(\mathcal{H})$ if

$$G_{s+t} = G_s \theta_s(G_t), \quad \forall s, t \geq 0, \quad G_0 = I.$$

A cocycle G is said to be *local* if G_t is in the commutant $(\theta_t(\mathcal{B}(\mathcal{H})))'$ for all t, that is, if G_t commutes with $\theta_t(Z)$ for all Z. It is said to be positive (respectively unitary, isometric, contractive) if each G_t is positive (respectively unitary, isometric, contractive).

The notion of local cocycles was introduced by Powers. It is closely connected with the notion of adaptedness of processes (see Section 9). As strong, weak operator topologies are better behaved on bounded sets we usually restrict ourselves to contractive cocycles. It is a simple exercise to see that contractive local cocycles form a semigroup, that is, if $\{G_t\}$, $\{H_t\}$ are contractive local cocycles then so is $\{G_t H_t\}$.

Right cocycles are defined analogus to left cocycles, just that we have $G_{s+t} = \theta_s(G_t)G_s$ for all s, t. Unless we mention otherwise by a cocycle we mean a left cocycle.

Definition 2.4: Two E_0-semigroups θ, θ' are said to be *cocycle conjugate* if there exists a third E_0-semigroup θ'' acting on $\mathcal{B}(\mathcal{H})$ and a unitary cocycle $U = \{U_t : t \geq 0\}$ of θ such that:

(i) $\theta_t''(Z) = U_t \theta_t(Z) U_t^*, \quad \forall Z \in \mathcal{B}(\mathcal{H}), t \geq 0;$

(ii) θ'' is conjugate to θ'.

If further U can be chosen such that $U_t u = u$ for all $u \in \mathcal{H}_0, t \geq 0$, and θ'', θ' are conjugate as dilations of τ, τ' then θ, θ' are said to be *cocycle conjugate as dilations* of τ, τ'.

It is to be noted that in Definition 2.4 as soon as U_t fixes every vector in $\mathcal{H}_0, \theta''$ is a dilation of τ. This is clear as we have $U_t P = P U_t = P$, where P

denotes the projection of $\mathcal{H}$ onto $\mathcal{H}_0$ and then

$$P\theta_t''(X)P = PU_t\theta_t(X)U_t^*P = P\theta_t(X)P = \tau_t(X) \quad \text{for } X \in \mathcal{B}(\mathcal{H}_0), t \geq 0.$$

The essence of Definition 2.4 is that if θ, θ' are acting on $\mathcal{B}(\mathcal{H})$ and are dilations of τ, they are cocycle conjugate as dilations if we can choose a cocycle which fixes all vectors in the initial space $\mathcal{H}_0$. The situation is further simplified by the following useful theorem. This result has been claimed earlier by W. Arveson (private communication) in the context of dilations of quantum dynamical semigroups on matrix algebras.

Theorem 2.5: Suppose θ, θ' are primary dilations of unital quantum dynamical semigroups τ, τ' respectively. Then θ, θ' are cocycle conjugate as dilations of τ, τ' iff they are conjugate as dilations of τ, τ'.

Proof: Suppose θ, θ' are cocycle conjugate as dilations. Let $\theta'', \{U_t : t \geq 0\}$ be as in Definition 2.4. Clearly it suffices to show that θ, θ'' are conjugate as dilations of τ, τ. Let P be the projection of $\mathcal{H}$ onto $\mathcal{H}_0$. For $t \geq 0$, take $\mathcal{H}_{t]} = $ range $\theta_t(P), \mathcal{H}_{t]}'' = $ range $\theta_t''(P)$ As θ, θ' are primary, we have $\mathcal{H}_{t]} \uparrow \mathcal{H}, \mathcal{H}_{t]}'' \uparrow \mathcal{H}$ as t increases to ∞. It is clear that U_t maps $\mathcal{H}_{t]}$ unitarily to $\mathcal{H}_{t]}''$. Define $V : \mathcal{H} \to \mathcal{H}$ by setting

$$Vz = U_t z \quad \text{for } z \in \mathcal{H}_{t]}, t \geq 0.$$

Now V is well-defined, for suppose $z \in \mathcal{H}_{s]}, 0 \leq s \leq t$, using the cocycle property

$$U_t z = U_t\theta_s(P)z = U_s\theta_s(U_{t-s})\theta_s(P)z = U_s\theta_s(P)z = U_s z.$$

Then it is clear that V extends to a unitary map of $\mathcal{H}$ which fixes vectors in $\mathcal{H}_0$. It maybe noted that $V = \lim_{t \to \infty} U_t$, where the limit is taken in strong operator topology. We claim that

$$\theta_s(Z) = V^*\theta_s''(VZV^*)V \qquad \forall Z \in \mathcal{B}(\mathcal{H}), s \geq 0.$$

Fix any $t \geq 0$, and consider $Z \in \mathcal{B}(\mathcal{H}_{t]})$. Then

$$
\begin{aligned}
V^*\theta_s''(VZV^*)V &= V^*U_s\theta_s(VZV^*)U_s^*V \\
&= V^*U_s\theta_s(U_tZU_t^*)U_s^*V \\
&= V^*U_{s+t}\theta_s(Z)U_{s+t}^*V \\
&= V^*U_{s+t}\theta_s(\theta_t(P)Z\theta_t(P))U_{s+t}^*V \\
&= V^*U_{s+t}\theta_{s+t}(P)\theta_s(Z)\theta_{s+t}(P)U_{s+t}^*V \\
&= V^*V\theta_{s+t}(P)\theta_s(Z)\theta_{s+t}(P)V^*V \\
&= \theta_s(Z).
\end{aligned}
$$

The equality extends to whole of $\mathcal{B}(\mathcal{H})$, due to normality of θ, θ''. This shows that θ, θ'' are conjugate as dilations. The converse statement of the theorem is trivial.   ∎

Here we briefly recall classification scheme of Powers and Arveson for E_0-semigroups of $\mathcal{B}(\mathcal{H})$, when $\mathcal{H}$ is separable. CCR flows (to be defined in Section 7) are most well-understood E_0-semigroups. Semigroups of automorphisms of $\mathcal{B}(\mathcal{H})$ and E_0-semigroups cocycle conjugate to CCR flows are said to be of type I. Units are certain intertwining operator semigroups (see Section 6) for E_0-semigroups. Those E_0-semigroups which have units are called spatial. All type I flows (for brevity we sometimes call E_0-semigroups as flows) are spatial, in fact they are completely spatial as they have plenty of units in a certain sense. E_0-semigroups which have units but are not of type I are said to be of type II. The rest, namely flows with no units what

so ever are said to be of type III. Arveson index of type III E_0-semigroups
is c (cardinality of real numbers) by definition. It is dimension of a certain
separable Hilbert space constructed out of units [Ar1] in type I and type II
case. Index and type are cocycle conjugacy invariants for E_0-semigroups.
An E_0-semigroup of type J (J=I, II, or III) and index n will be called a
type J_n E_0-semigroup. Type I semigroups of non-zero index are completely
classified (up to cocycle conjugacy) by their index, natural representatives
for these equivalence classes being CCR flows (or CAR flows) of desired index. Type I semigroups of index zero are just semigroups of automorphisms
and we have one equivalence class for each possible dimension of $\mathcal{H}$. Of
course, when $\mathcal{H}$ is finite dimensional all E_0-semigroups are semigroups of
automorphisms.

For a unital quantum dynamical semigroup we define index (respectively
type) as the index (respectively type) of its minimal dilation. W. Arveson
[Ar4] has come up with a more intrinsic definition of index for quantum
dynamical semigroups. Neverthless it is designed in such a way that it
matches with our definition. An E_0-semigroup dilation is said to be a type J
(J= I, II, III) dilation if it is of type J as an E_0-semigroup under classification
described above. The following result of Arveson is well-known. The only
reason for its inclusion here is that we need it quite frequently and therefore
it is convenient to have a reference point.

Proposition 2.6 (Arveson) : Let $\mathcal{H}, \mathcal{K}$ be complex, separable Hilbert
spaces where $\mathcal{H}$ is infinite dimensional. Suppose θ is an E_0-semigroup
of $\mathcal{B}(\mathcal{H})$ and α is a semigroup of automorphisms of $\mathcal{B}(\mathcal{K})$. Then the E_0-

semigroup $\alpha \otimes \theta$ of $\mathcal{B}(\mathcal{K} \otimes \mathcal{H})$ is cocycle conjugate to θ.

Proof: This follows easily from Arveson's result that two E_0-semigroups are cocycle conjugate iff their product systems are isomorphic. See the Corollary on page 33 of [Ar1] (The term 'outer conjugacy' used there is an old terminology for cocycle conjugacy) or Theorem 2.10 of [Po4].   ■

3 Minimal dilation and induced semigroup

We are going to study a bit as to how does the minimal dilation sits inside a general dilation. The set up we have here and the various constructions we are going to make (including the notation) will be used through-out the article. Let $\theta = \{\theta_t : t \geq 0\}$ be an E_0-semigroup of $\mathcal{B}(\mathcal{H})$, where $\mathcal{H}$ is a complex separable Hilbert space with its inner-product $(\langle \cdot, \cdot \rangle)$ anti-linear in the first-varible. For x, y in $\mathcal{H}$ the operator which maps $z \in \mathcal{H}$ to $\langle y, z \rangle x$ is denoted by $|x\rangle\langle y|$. Let $\mathcal{H}_0$ be a closed subspace of $\mathcal{H}$ such that the orthogonal projection P of $\mathcal{H}$ on to $\mathcal{H}_0$ is increasing for θ. Let $\tau = \{\tau_t : t \geq 0\}$ be the compression of θ to $\mathcal{B}(\mathcal{H}_0)$, defined by (2.1). So θ is a dilation of τ.

For $t \geq 0$, take $\mathcal{H}_{t]} = $ range $(\theta_t(P))$. By definition θ is a primary dilation if $\mathcal{H}_{t]} \uparrow \mathcal{H}$ as $t \uparrow \infty$. For any quantum dynamical semigroup α of $\mathcal{B}(\mathcal{H})$ we find it convenient to denote the operator $\alpha_{r_1}(X_1) \cdots \alpha_{r_n}(X_n)$ by $\alpha(\underline{r}, \underline{X})$ for n-tuples $\underline{r} = (r_1, \cdots, r_n), r_i \geq 0$, and $\underline{X} = (X_1, \cdots, X_n), X_i \in \mathcal{B}(\mathcal{H})$. With this convention define $\hat{\mathcal{H}}, \hat{\mathcal{H}}_{t]}$ by

$$\hat{\mathcal{H}} = \overline{\text{span}} \{\theta(\underline{r}, \underline{X})u : r_i \geq 0, X_i \in \mathcal{B}(\mathcal{H}_0), 1 \leq i \leq n, u \in \mathcal{H}_0, n \geq 0\}$$

$$\hat{\mathcal{H}}_{t]} = \overline{\text{span}} \{\theta(\underline{r}, \underline{X})u : t \geq r_i \geq 0, X_i \in \mathcal{B}(\mathcal{H}_0), 1 \leq i \leq n, u \in \mathcal{H}_0, n \geq 0\}$$

(Here and elsewhere $\theta(\underline{r}, \underline{X})u$ is interpreted as vector u, if $n = 0$). It is clear that $\hat{\mathcal{H}}_{t]}$ increases to $\hat{\mathcal{H}}$ as t increases to ∞. θ is the minimal dilation of τ if and only if $\hat{\mathcal{H}} = \mathcal{H}$. Usually we call $\mathcal{H}_0$ as *initial space*, $\mathcal{H}$ as *dilation space*, and $\hat{\mathcal{H}}$ as *minimal dilation space*.

Note that for $r_2 \geq r_1 \geq r_3 \geq 0$, and $X_i \in \mathcal{B}(\mathcal{H}_0)$

$$
\begin{aligned}
\theta_{r_1}(X_1)\theta_{r_2}(X_2)\theta_{r_3}(X_3) &= \theta_{r_1}(X_1\theta_{r_2-r_1}(X_2))\theta_{r_3}(X_3) \\
&= \theta_{r_1}(X_1\theta_{r_2-r_1}(X_2))\theta_{r_3}(P)\theta_{r_3}(X_3) \\
&= \theta_{r_1}(X_1\tau_{r_2-r_1}(X_2))\theta_{r_3}(X_3).
\end{aligned}
$$

Similarly for $r_2 \geq r_3 \geq r_1$,

$$
\theta_{r_1}(X_1)\theta_{r_2}(X_2)\theta_{r_3}(X_3) = \theta_{r_1}(X_1)\theta_{r_3}(\tau_{r_2-r_1}(X_2)X_3).
$$

It follows that in defining $\hat{\mathcal{H}}, \hat{\mathcal{H}}_{t]}$ we can assume that $r_1, r_2, \cdots, r_n$ are in decreasing order, that is $\hat{\mathcal{H}} = \overline{\text{span}} \, \mathcal{M}, \hat{\mathcal{H}}_{t]} = \overline{\text{span}} \, \mathcal{M}_{t]}$, where

$$
\mathcal{M} = \{\theta(\underline{r}, \underline{X})u : (\underline{r}, \underline{X}, u) \in \mathcal{N}\},
$$

$$
\mathcal{M}_{t]} = \{\theta(\underline{r}, \underline{X})u : (\underline{r}, \underline{X}, u) \in \mathcal{N}_t\},
$$

$$
\mathcal{N} = \{(\underline{r}, \underline{X}, u) : r_1 \geq r_2 \geq \cdots \geq r_n \geq 0, X_i \in \mathcal{B}(\mathcal{H}_0), u \in \mathcal{H}_0, n \geq 0\},
$$

$$
\mathcal{N}_t = \{(\underline{r}, \underline{X}, u) : t = r_1 \geq r_2 \geq \cdots \geq r_n \geq 0, X_i \in \mathcal{B}(\mathcal{H}_0), u \in \mathcal{H}_0, n \geq 1\}.
$$

For any Hilbert space $\mathcal{K}$ we denote its identity operator by $1_\mathcal{K}$. If $\mathcal{K}$ is a subspace of $\mathcal{H}$, $1_\mathcal{K}$ is identified with the projection of $\mathcal{H}$ onto $\mathcal{K}$ and more generally any operator $A \in \mathcal{B}(\mathcal{K})$ is identified with its natural embedding $1_\mathcal{K} A 1_\mathcal{K}$ in $\mathcal{B}(\mathcal{H})$. For $\xi_1, \xi_2 \in \mathcal{M}_{s]}, \xi_3 \in \mathcal{M}_{t]}$ clearly $\theta_t(|\xi_1\rangle\langle\xi_2|)\xi_3 \in \mathcal{M}_{s+t]}$. A closer look shows that for every $s, t, \theta_t(1_{\hat{\mathcal{H}}_{s]}}) \geq 1_{\hat{\mathcal{H}}_{s+t]}}$. Hence $\theta_t(1_{\hat{\mathcal{H}}}) \geq \theta_t(1_{\hat{\mathcal{H}}_{s]}}) \geq 1_{\hat{\mathcal{H}}_{s+t]}}$. Fixing t, and taking limit as $s \to \infty$, we obtain $\theta_t(1_{\hat{\mathcal{H}}}) \geq 1_{\hat{\mathcal{H}}}$. In other words $1_{\hat{\mathcal{H}}}$ is an increasing projection for θ. Let $\hat{\tau} = \{\hat{\tau}_t : t \geq 0\}$ be the compression of θ to $\mathcal{B}(\hat{\mathcal{H}})$. Clearly $\theta_t(|\xi\rangle\langle\eta|)(\hat{\mathcal{H}}) \subseteq \hat{\mathcal{H}}$, for $\xi, \eta \in \mathcal{M}$. Then by normality of θ_t it is clear that $\theta_t(Y)\hat{\mathcal{H}} \subseteq \hat{\mathcal{H}}$, for every $Y \in \mathcal{B}(\hat{\mathcal{H}})$. Therefore $\hat{\mathcal{H}}$ reduces θ_t, that is,

$$
\hat{\tau}(\underline{r}, \underline{X})u = \theta(\underline{r}, \underline{X})u, \tag{3.1}
$$

for $(\underline{r}, \underline{X}, u)$ in $\mathcal{N}$. Further for $X \in \mathcal{B}(\mathcal{H}_0), (\underline{r}, \underline{Y}, u), (\underline{r}, \underline{Z}, v) \in \mathcal{N}_t$,

$$\langle \theta(\underline{r}, \underline{Y})u, \theta_t(X)\theta(\underline{r}, \underline{Z})v \rangle$$
$$= \langle u, \tau_{r_n}(Y_n^* \cdots \tau_{r_2 - r_3}(Y_2^* \tau_{r_1 - r_2}(Y_1^* X Z_1)Z_2) \cdots Z_n)v \rangle \qquad (3.2)$$

(See [Bh3]).

Thanks to Arveson we have a way of associating continuous tensor product systems of Hilbert spaces with E_0-semigroups. Here we have two of them $\hat{\tau}, \theta$. Product systems of $\hat{\tau}$ and θ are related. In fact, we can say that the product system of $\hat{\tau}$ is a sub-product system of that of θ, in a very precise way. We explain the basic ideas involved without going in to the technicalities of defining a product system [Ar1]. We take the approach of [Bh2].

Fix a unit vector $a \in \mathcal{H}_0$. Now for $t \geq 0$ consider the following subspaces of $\mathcal{H}$:

$$\mathcal{P}_t = \text{ range } \theta_t(|a\rangle\langle a|),$$
$$\hat{\mathcal{P}}_t = \text{ range } \hat{\tau}_t(|a\rangle\langle a|).$$

Note that in view of (3.1) $\hat{\mathcal{P}}_t$ is a subspace of $\mathcal{P}_t$. We call a as *ground vector*. Of course, $\mathcal{P}_t, \hat{\mathcal{P}}_t$ depend upon a but we are suppressing it by fixing a. Most of the constructions we are going to make will not depend upon the choice of a upto unitary equivalence. Define $W_t : \mathcal{H} \otimes \mathcal{P}_t \to \mathcal{H}$ by

$$W_t(z \otimes \theta_t(|a\rangle\langle a|)y) = \theta_t(|z\rangle\langle a|)y, \quad z, y \in \mathcal{H}.$$

Proposition 3.1: For every t, W_t is a unitary operator such that for all s

(i) $W_t(\mathcal{H}_{s]} \otimes \mathcal{P}_t) = \mathcal{H}_{s+t]}$, (ii) $W_t(\mathcal{H} \otimes \mathcal{P}_t) = \mathcal{H}$, (iii) $W_t(\hat{\mathcal{H}}_{s]} \otimes \hat{\mathcal{P}}_t) = \hat{\mathcal{H}}_{s+t]}$,

(iv) $W_t(\hat{\mathcal{H}} \otimes \hat{\mathcal{P}}_t) = \hat{\mathcal{H}}$, (v) $W_t(\mathcal{P}_s \otimes \mathcal{P}_t) = \mathcal{P}_{s+t}$, (vi) $W_t(\hat{\mathcal{P}}_s \otimes \hat{\mathcal{P}}_t) = \hat{\mathcal{P}}_{s+t}$.

Furthermore, (vii) $\theta_t(Z) = W_t(Z \otimes 1_{\mathcal{P}_t})W_t^*$ for $Z \in \mathcal{B}(\mathcal{H})$, (viii) $\hat{\tau}_t(X) = W_t(X \otimes 1_{\hat{\mathcal{P}}_t})W_t^*$ for $X \in \mathcal{B}(\hat{\mathcal{H}})$.

Proof: By direct computation of inner products it is seen that W_t is isometric. Unitarity will follow from (ii). W_t maps $\mathcal{H}_{s]} \otimes \mathcal{P}_t$ into $\mathcal{H}_{s+t]}$ is obvious. Now if $\{e_k\}_{k \geq 1}$ is an ortho-normal basis of $\mathcal{H}$, for any k we have $\theta_s(P)e_k \in \mathcal{H}_{s]}$ and $\theta_t(|a\rangle\langle e_k|)y = \theta_t(|a\rangle\langle a|)\theta_t(|a\rangle\langle e_k|)y \in \mathcal{P}_t$ for $y \in \mathcal{H}$. And then using normality of θ_t

$$
\begin{aligned}
W_t \sum_k \theta_s(P)e_k \otimes \theta_t(|a\rangle\langle e_k|)y &= \sum_k \theta_t(|\theta_s(P)e_k\rangle\langle e_k|)y \\
&= \theta_{s+t}(P)(\sum_k \theta_t(|e_k\rangle\langle e_k|)y) \\
&= \theta_{s+t}(P)y.
\end{aligned}
$$

This proves (i). Statements (ii) to (vi) are proved in a similar fashion. For $Z \in \mathcal{B}(\mathcal{H}), z, y \in \mathcal{H}$ we have

$$
\begin{aligned}
\theta_t(Z)W_t(z \otimes \theta_t(|a\rangle\langle a|)y) &= \theta_t(Z)\theta_t(|z\rangle\langle a|)y \\
&= \theta_t(|Zz\rangle\langle a|)y \\
&= W_t(Z \otimes 1_{\mathcal{P}_t})(z \otimes \theta_t(|a\rangle\langle a|)y).
\end{aligned}
$$

Hence $\theta_t(Z)W_t = W_t(Z \otimes 1_{\mathcal{P}_t})$ or $\theta_t(Z) = W_t(Z \otimes 1_{\mathcal{P}_t})W_t^*$. On making use of (3.1) a similar argument gives (viii). ∎

Any unital representation of $\mathcal{B}(\mathcal{H})$ is just the identity representation with some multiplicity. In Proposition 3.1 we have made this explicit for representations θ_t and $\hat{\tau}_t$. Let $U_{s,t}$ denote the unitary operator obtained by restricting W_t to domain $\mathcal{P}_s \otimes \mathcal{P}_t$ and range $\mathcal{P}_{s+t}$.

Proposition 3.2: For $r, s, t \geq 0$, (i) $W_s(W_t \otimes 1_{\mathcal{P}_s}) = W_{s+t}(1_{\mathcal{H}} \otimes U_{t,s})$, (ii) $U_{r,s+t}(1_{\mathcal{P}_r} \otimes U_{s,t}) = U_{r+s,t}(U_{r,s} \otimes 1_{\mathcal{P}_t})$.

Proof: For x, y, z in $\mathcal{H}$,

$$W_s(W_t \otimes 1_{\mathcal{P}_s})(z \otimes \theta_t(|a\rangle\langle a|)y \otimes \theta_s(|a\rangle\langle a|)x)$$
$$= W_s(\theta_t(|z\rangle\langle a|)y \otimes \theta_s(|a\rangle\langle a|)x)$$
$$= \theta_s(|\theta_t(|z\rangle\langle a|)y\rangle\langle a|)x$$
$$= \theta_{s+t}(|z\rangle\langle a|)\theta_s(|y\rangle\langle a|)x$$
$$= W_{s+t}(z \otimes \theta_{s+t}(|a\rangle\langle a|)\theta_s(|y\rangle\langle a|)x)$$
$$= W_{s+t}(1_{\mathcal{H}} \otimes U_{t,s})(z \otimes \theta_t(|a\rangle\langle a|)y \otimes \theta_s(|a\rangle\langle a|)x).$$

This shows (i). Proof of (ii) is analogus.   ■

The family of Hilbert spaces $\{\mathcal{P}_t : t \geq 0\}$ with unitary operators $\{U_{s,t} : s, t \geq 0\}$ forms a continuous tensor product sytem. Apart from some technical conditions [Ar1] this just means that $\mathcal{P}_{s+t}$ is isomorphic to $\mathcal{P}_s \otimes \mathcal{P}_t$ through unitaries $U_{s,t}$ and the associativety condition, that is (ii) of Proposition 3.2, holds. Taking $\hat{U}_{s,t}$ as $U_{s,t}$ restricted to domain $\hat{\mathcal{P}}_s \otimes \hat{\mathcal{P}}_t$ and range $\hat{\mathcal{P}}_{s+t}$, we see that $\{\hat{\mathcal{P}}_t : t \geq 0\}$ with $\{\hat{U}_{s,t} : s, t \geq)\}$ is another product sytem. This is the product system associated with $\hat{\tau}$. Now it should be clear as to in what sense we mean $(\hat{\mathcal{P}}_t, \hat{U}_{s,t})$ to be a sub-product system of $(\mathcal{P}_t, U_{s+t})$.

Apart from the minimal dilation $\hat{\tau}$ there is another semigroup of interest we can form from the pair (τ, θ). We call it the *induced semigroup* of τ (induced by dilation θ) and denote it by $\tilde{\tau}$. Define $\tilde{\tau} = \{\tilde{\tau}_t : t \geq 0\}, \tilde{\tau}_t : \mathcal{B}(\mathcal{H}) \to \mathcal{B}(\mathcal{H})$

by

$$\tilde{\tau}_t(Z) = W_t(Z \otimes 1_{\hat{\mathcal{P}}_t})W_t^* \quad \text{for } Z \in \mathcal{B}(\mathcal{H}). \tag{3.3}$$

This definition is to be compared with (vii), (viii) of Proposition 3.1.

Theorem 3.3: Consider $\tilde{\tau}$ defined as above. Then $\tilde{\tau}$ is an e_0-semigroup of $\mathcal{B}(\mathcal{H})$. It is independent of the choice of ground vector a. Moreover $\tau, \hat{\tau}$ are compressions of $\tilde{\tau}$. For every t, $\theta_t - \tilde{\tau}_t$ is completely positive.

Proof: Clearly for any t, $\tilde{\tau}_t$ is a normal $*$-endomorphism of $\mathcal{B}(\mathcal{H})$. Now for $Z \in \mathcal{B}(\mathcal{H})$ and $s, t \geq 0$,

$$
\begin{aligned}
\tilde{\tau}_s(\tilde{\tau}_t(Z)) &= W_s(W_t(Z \otimes 1_{\hat{\mathcal{P}}_t})W_t{}^* \otimes 1_{\hat{\mathcal{P}}_s})W_s{}^* \\
&= W_s(W_t \otimes 1_{\mathcal{P}_s})(Z \otimes 1_{\hat{\mathcal{P}}_t} \otimes 1_{\hat{\mathcal{P}}_s})(W_t{}^* \otimes 1_{\mathcal{P}_s})W_s{}^* \\
&= W_{s+t}(1 \otimes U_{t,s})(Z \otimes 1_{\hat{\mathcal{P}}_t} \otimes 1_{\hat{\mathcal{P}}_s})(1 \otimes U_{t,s}^*)W_{s+t}^* \\
&= W_{s+t}(Z \otimes 1_{\hat{\mathcal{P}}_{s+t}})W_{s+t}^* \\
&= \hat{\tau}_{s+t}(Z).
\end{aligned}
$$

Note that as $\tilde{\tau}_t$ is normal, linear it is determined by its action on rank one operators. For $x, y \in \mathcal{H}$ and any unit vector $b \in \mathcal{H}_0$ we have

$$
\begin{aligned}
\tilde{\tau}_t(|x\rangle\langle y|) &= W_t(|x\rangle\langle y| \otimes 1_{\hat{\mathcal{P}}_t})W_t^* \\
&= W_t(|x\rangle\langle b| \otimes 1_{\mathcal{P}_t})W_t^*.W_t(|b\rangle\langle b| \otimes 1_{\hat{\mathcal{P}}_t})W_t^*.W_t(|b\rangle\langle y| \otimes 1_{\mathcal{P}_t})W_t^* \\
&= \theta_t(|x\rangle\langle b|)\hat{\tau}_t(|b\rangle\langle b|)\theta_t(|b\rangle\langle y|). \tag{3.4}
\end{aligned}
$$

Clearly the last expression is independent of the choice of ground vector a. It also shows that $t \mapsto \tilde{\tau}_t(Z)$ is weakly measurable and hence strongly continuous for any $Z \in \mathcal{B}(\mathcal{H})$ (See Appendix A). For $Y \in \mathcal{B}(\hat{\mathcal{H}})$ from the very definition of $\tilde{\tau}$ (and (viii) of Proposition 3.1) $\tilde{\tau}_t(Y) = \hat{\tau}_t(Y)$. Hence the compression of $\tilde{\tau}$ to $\mathcal{B}(\hat{\mathcal{H}})$ is $\hat{\tau}$. As the compression of $\hat{\tau}$ to $\mathcal{B}(\mathcal{H}_0)$ is τ we get the same result for $\tilde{\tau}$ as well. Finally observe that for $Z \in \mathcal{B}(\mathcal{H})$

$$\theta_t(Z) - \tilde{\tau}_t(Z) = W_t(Z \otimes (1_{\mathcal{P}_t} - 1_{\hat{\mathcal{P}}_t}))W_t{}^*, \quad t \geq 0.$$

As $\hat{\mathcal{P}}_t$ is a subspace of $\mathcal{P}_t, \theta_t - \tilde{\tau}_t$ is completely positive.   ■

Remark 3.4: θ is the minimal dilation of τ iff θ is primary and $\theta = \tilde{\tau}$.

Proof: Suppose θ is primary and $\theta = \tilde{\tau}$. From the definition of $\tilde{\tau}$ we obtain $\hat{\mathcal{P}}_t = \mathcal{P}_t$. Hence $\hat{\mathcal{H}}_{t]} = W_t(\mathcal{H}_0 \otimes \hat{\mathcal{P}}_t) = W_t(\mathcal{H}_0 \otimes \mathcal{P}_t) = \mathcal{H}_{t]} \uparrow \mathcal{H}$. Therefore $\hat{\mathcal{H}} = \mathcal{H}$, and θ is minimal. Conversely if θ is the minimal dilation of τ, we know that θ is primary, also $\hat{\mathcal{H}}_{t]} = \text{range } \theta_t(P)$, implying $\hat{\mathcal{P}}_t = \mathcal{P}_t$, hence $\tilde{\tau}_t = \theta_t$.   ■

By considering examples such as $\theta = \alpha \otimes \hat{\tau}$, with $\alpha_t(X) \equiv X$, one can see that θ is needed to be primary in this Remark.

On constructing the product system of $\tilde{\tau}$, using a unit vector $a \in \mathcal{H}_0$, we see that $\tilde{\tau}$ and $\hat{\tau}$ have same product systems. This leads to the following result.

Proposition 3.5: (i) If τ is elementary then $\hat{\tau}, \tilde{\tau}$ are also elementary. (ii) If τ is not a semigroup of automorphisms on a finite dimensional space, then $\hat{\tau}$ and $\tilde{\tau}$ are cocycle conjugate through a cocycle of isometries.

Proof: (i) If τ is elementary then so is $\hat{\tau}$ as it is a semigroup of automorphisms. Indeed if $\tau_t(\cdot) = A_t(\cdot)A_t^*$ for a contraction semigroup $\{A_t : t \geq 0\}$ with $A_{t_0} \neq 0$ for some $t_0 > 0$ then $\hat{\tau}_t(\cdot) = \hat{A}_t(\cdot)\hat{A}_t{}^*$, where $\{\hat{A}_t : t \geq 0\}$

is Sz. Nagy's minimal isometric dilation [SzF] of $\{A_t\}$ (see Theorem 2.6 of [Bh1] for details). In the present case, as τ, $\hat{\tau}$ are unital $\{A_t : t \geq 0\}$ are co-isometries and $\{\hat{A}_t : t \geq 0\}$ are unitaries. Fix any unit vector b in $\mathcal{H}_0$. Consider $\tilde{A}_t : \mathcal{H} \to \mathcal{H}$ defined by

$$\tilde{A}_t x = \theta_t(|x\rangle\langle b|)\hat{A}_t b, \quad \text{for} \quad x \in \mathcal{H}.$$

It is easily verified that $\{\tilde{A}_t : t \geq 0\}$ are isometric. For $x, y \in \mathcal{H}$

$$\tilde{A}_t|x\rangle\langle y|\tilde{A}_t^* = \theta_t(|x\rangle\langle b|)\hat{\tau}_t(|b\rangle\langle b|)\theta_t(|b\rangle\langle y|),$$

and hence from (3.4), $\tilde{\tau}_t(Z) = \tilde{A}_t Z \tilde{A}_t^*, \forall Z \in \mathcal{B}(\mathcal{H})$. Now for $s, t \geq 0$, $y \in \mathcal{H}$

$$
\begin{aligned}
\tilde{A}_s \tilde{A}_t y &= \theta_s(|\theta_t(|y\rangle\langle b|)\hat{A}_t b\rangle\langle b|)\hat{A}_s b = \theta_{s+t}(|y\rangle\langle b|)\theta_s(|\hat{A}_t b\rangle\langle b|)\hat{A}_s b \\
&= \theta_{s+t}(|y\rangle\langle b|)\hat{\tau}_s(|\hat{A}_t b\rangle\langle b|)\hat{A}_s b = \theta_{s+t}(|y\rangle\langle b|)\hat{A}_{s+t} b \\
&= \tilde{A}_{s+t} y.
\end{aligned}
$$

Hence $\tilde{\tau}$ is elementary. (We have not verified strong operator continuity of $t \mapsto \tilde{A}_t$, however it follows easily from Appendix A, as $\tilde{A}_t$ becomes a unit of θ.)

(ii) The assumption on τ means that $\hat{\mathcal{H}}$ is infinite dimensional. Let $L : \mathcal{H} \to \mathcal{H}$ be an isometry with range $L = \hat{\mathcal{H}}$. Then L^* restricted to $\hat{\mathcal{H}}$ is a unitary mapping $\hat{\mathcal{H}}$ to $\mathcal{H}$. Consider the quantum dynamical semigroup $\bar{\tau}$ on $\mathcal{B}(\mathcal{H})$ defined by

$$\bar{\tau}_t(Z) = L^*\hat{\tau}_t(LZL^*)L \quad Z \in \mathcal{B}(\mathcal{H}), t \geq 0.$$

This way $\hat{\tau}$ and $\bar{\tau}$ are conjugate. Also note that as $LL^* = 1_{\hat{\mathcal{H}}}, \hat{\tau}_t(LL^*) = \hat{\tau}_t(LL^*) = LL^*$. For $t \geq 0$ take $V_t = \tilde{\tau}_t(L^*)L$. Now for $s, t \geq 0$

$$V_s \bar{\tau}_s(V_t) = \tilde{\tau}_s(L^*)LL^*\hat{\tau}_s(LV_t L^*)L = \tilde{\tau}_s(L^*)\hat{\tau}_s(LV_t L^*)L$$

$$= \tilde{\tau}_s(L^*)\tilde{\tau}_s(LV_tL^*)L = \tilde{\tau}_s(V_tL^*)L = \tilde{\tau}_s(\tilde{\tau}_t(L^*)LL^*)L$$

$$= \tilde{\tau}_{s+t}(L^*)L = V_{s+t},$$

and we also have

$$V_t\bar{\tau}_t(Z)V_t{}^* = V_tL^*\hat{\tau}_t(LZL^*)LV_t{}^* = \tilde{\tau}_t(L^*)LL^*\tilde{\tau}_t(LZL^*)LL^*\tilde{\tau}_tLL^*$$

$$= \tilde{\tau}_t(L^*)\tilde{\tau}_t(LZL^*)\tilde{\tau}_t(L^*) = \tilde{\tau}_t(Z). \qquad (3.5)$$

Further $V_t{}^*V_t = L^*\tilde{\tau}_t(LL^*)L = L^*LL^*L = 1$. The strong (operator) continuity of $t \mapsto V_t$ is clear as $t \mapsto \tilde{\tau}_t(L^*)$ is continuous strongly from Theorem
3.3. Hence $\tilde{\tau}$ and $\hat{\tau}$ are cocycle conjugate through isometric cocycle $\{V_t\}$. ∎

In the sequel more often than not we will be assuming that the dilation
under consideration is primary. This is not so bad because all dilations
can be described through primary dilations and this is possible essentially
because (in the words of Arveson) there is no direct sum operation for E_0-
semigroups.

Theorem 3.6 : Let τ, θ be as above. Suppose θ is not a primary dilation
of τ.

(i) If θ is a semigroup of automorphisms then decomposing $\mathcal{H}$ as $\mathcal{H} = \hat{\mathcal{H}} \oplus
(\hat{\mathcal{H}})^\perp$, there are semigroups of unitaries $\{A_t\}, \{B_t\}$ of $\hat{\mathcal{H}}, (\hat{\mathcal{H}})^\perp$ respectively
such that $\hat{\tau}_t(X) = A_tXA_t^*$, for $X \in \mathcal{B}(\hat{\mathcal{H}})$ and $\theta_t(Z) = (A_t \oplus B_t)Z(A_t^* \oplus B_t^*)$,
for $Z \in \mathcal{B}(\mathcal{H})$;

(ii) If θ is not a semigroup of automorphisms then $\mathcal{H}$ factorizes as $\mathcal{H} = \mathbb{C}^2 \otimes \mathcal{K}$,
for some Hilbert space $\mathcal{K}$ with $\mathcal{H}_0 \subseteq \hat{\mathcal{H}} \subseteq (\binom{1}{0} \otimes \mathcal{K})$, and a primary dilation

ψ of τ in $\mathcal{B}(\mathcal{K})$, with a unitary cocycle $\{U_t\}$ of ψ such that

$$\theta_t\left(\begin{bmatrix} X & Y \\ Z & W \end{bmatrix}\right) = \begin{bmatrix} \psi_t(X) & \psi_t(Y)U_t^* \\ U_t\psi_t(Z) & U_t\psi_t(W)U_t^* \end{bmatrix}$$

for $X, Y, Z, W \in \mathcal{B}(\mathcal{K}), t \geq 0$.

Proof : (i) This is fairly obvious once we observe that if $\theta_t(Z) = U_t Z U_t^*$ for a unitary semigroup $\{U_t\}$ then U_t, U_t^* must leave $\hat{\mathcal{H}}$ invariant.

(ii) Taking limit in strong operator topology $Q = \lim_{t\to\infty} \theta_t(P)$, and we have $0 < Q < I$. Clearly $\theta_t(Q) = Q$ and $\theta_t(1 - Q) = (1 - Q)$. Take $\mathcal{K} = \text{range } Q$. If θ is not a semigroup of automorphisms then for any rank one projection $|x\rangle\langle x|, \theta_t(|x\rangle\langle x|)$ is of infinite rank for $t > 0$. Hence $\dim \mathcal{K} = \dim \mathcal{K}^\perp = \infty$. So we can identify $\mathcal{H}$ with $\mathbb{C}^2 \otimes \mathcal{K}$, in such a way that $\mathcal{K}$ gets identified with $\left(\begin{pmatrix} 1 \\ 0 \end{pmatrix}\right) \otimes \mathcal{K}$. Let ψ be the compression of θ to $\mathcal{B}(\mathcal{K})$. Clearly ψ is a primary dilation of τ. As $\theta_t(1 - Q) = (1 - Q)$, compression of θ to $\mathcal{B}(\mathcal{K}^\perp)$ is also an E_0-semigroup. Then it follows that (see [Ar7] for details).

$$\theta_t\left(\begin{bmatrix} 0 & 0 \\ 1 & 0 \end{bmatrix}\right) = \begin{bmatrix} 0 & 0 \\ U_t & 0 \end{bmatrix}$$

for some unitary cocycle $\{U_t\}$ of ψ, and (ii) is satisfied. ∎

In (ii) of Theorem 3.6, writing θ as

$$\theta_t(Z) = \begin{bmatrix} 1 & 0 \\ 0 & U_t \end{bmatrix} [(id \otimes \psi_t)(Z)] \begin{bmatrix} 1 & 0 \\ 0 & U_t^* \end{bmatrix},$$

from Proposition 2.6 we see that θ and ψ are cocycle conjugate.

We end this section with a curious result. It is clear that if P is increasing then $\theta_t(P)$ is increasing for any $t > 0$. For $t_0 > 0$, let $\tau^{(t_0)}$ be the compression of θ by $\theta_{t_0}(P)$. We call quantum dynamical semigroups $\tau^{(t_0)}, t_0 > 0$ (strict inequality is important) as *derived semigroups*.

Theorem 3.7: Suppose θ is a primary dilation then it is a minimal dilation for all derived semigroups.

Proof : Fix $t_0 > 0$. The derived semigroup $\tau^{(t_0)}$ acts on $\mathcal{B}(\mathcal{H}_{t_0]})$ by

$$\tau_t^{(t_0)}(Y) = \theta_{t_0}(P)\theta_t(Y)\theta_{t_0}(P), t \geq 0, Y \in \mathcal{B}(\mathcal{H}_{t_0]}).$$

Clearly θ is a primary dilation of $\tau^{(t_0)}$. It is minimal if

$$\overline{\text{span}}\{\theta(\underline{r},\underline{Y})y : r_i \geq 0, Y_i \in \mathcal{B}(\mathcal{H}_{t_0]}), y \in \mathcal{H}_{t_0]}\} = \mathcal{H}.$$

We derive this from,

$$\overline{\text{span}}\{\theta(\underline{r},\underline{Y})y : t \geq r_i \geq 0, Y_i \in \mathcal{B}(\mathcal{H}_{t_0]}), y \in \mathcal{H}_{t_0]}\} = \mathcal{H}_{t+t_0]}.$$

for $t > 0$. Here it is clear that $\theta(\underline{r},\underline{Y})y \in \mathcal{H}_{t+t_0]}$, as range $\theta_{r_1}(Y_1) \subseteq$ range $\theta_{r_1}(\theta_{t_0}(P)) = \mathcal{H}_{t+t_0]}$. To see that $\mathcal{H}_{t+t_0]}$ is actually spanned by such vectors we try to approximate an arbitrary vector in $\mathcal{H}_{t+t_0]}$. For $z \in \mathcal{H}$, we want to approximate $\theta_{t+t_0}(P)z$. As $\theta_{t+t_0}(P)z = \sum_k \theta_{t+t_0}(|u_k\rangle\langle u_k|)z$, for any ortho-normal basis $\{u_k\}_{k\geq 1}$ of $\mathcal{H}_0$, it is sufficient to consider vectors of the form $\theta_{t+t_0}(|x_1\rangle\langle x_2|)z$, $x_1, x_2 \in \mathcal{H}_0, z \in \mathcal{H}$. Let $n \geq 0$ be the integer such that $0 \leq t - nt_0 < t_0$. Let $\{e_k : k \geq 1\}$ be an ortho-normal basis of $\mathcal{H}$. As θ is normal, unital we obtain

$$\theta_{t+t_0}(|x_1\rangle\langle x_2|)z = \sum_{k_1,\cdots,k_n} \theta_{t+t_0}(|x_1\rangle\langle x_2|)\theta_t(|e_{k_1}\rangle\langle e_{k_1}|)\theta_{t-t_0}(|e_{k_2}\rangle\langle e_{k_2}|)$$
$$\cdots \theta_{t-nt_0}(|e_{k_n}\rangle\langle e_{k_n}|)z. \tag{3.6}$$

Now

$$\theta_{t+t_0}(|x_1\rangle\langle x_2|)\theta_t(|e_{k_1}\rangle\langle e_{k_1}|)\theta_{t-t_0}(|e_{k_2}\rangle\langle e_{k_2}|)\cdots\theta_{t-nt_0}(|e_{k_n}\rangle\langle e_{k_n}|)z$$

$$= \theta_t(|y_1\rangle\langle e_{k_1}|)\theta_{t-t_0}(|e_{k_2}\rangle\langle e_{k_2}|)\cdots\theta_{t-nt_0}(|e_{k_n}\rangle\langle e_{k_n}|)z$$

$$= \theta_t(|y_1\rangle\langle a|)\theta_t(|a\rangle\langle e_{k_1}|)\theta_{t-t_0}(|e_{k_2}\rangle\langle e_{k_2}|)\cdots\theta_{t-nt_0}(|e_{k_n}\rangle\langle e_{k_n}|)z,$$

where $y_1 = \theta_{t_0}(|x_1\rangle\langle x_2|)e_{k_1} \in \mathcal{H}_{t_0]}$, and a is a fixed unit vector in $\mathcal{H}_0$. Further

$$\theta_t(|y_1\rangle\langle a|)\theta_t(|a\rangle\langle e_{k_1}|)\theta_{t-t_0}(|e_{k_2}\rangle\langle e_{k_2}|)\cdots\theta_{t-nt_0}(|e_{k_n}\rangle\langle e_{k_n}|)z$$

$$= \theta_t(|y_1\rangle\langle a|)\theta_{t-t_0}(|y_2\rangle\langle e_{k_2}|)\theta_{t-2t_0}(|e_{k_3}\rangle\langle e_{k_3}|)\cdots\theta_{t-nt_0}(|e_{k_n}\rangle\langle e_{k_n}|)z$$

$$= \theta_t(|y_1\rangle\langle a|)\theta_{t-t_0}(|y_2\rangle\langle a|)\theta_{t-t_0}(|a\rangle\langle e_{k_2}|)\cdots\theta_{t-nt_0}(|e_{k_n}\rangle\langle e_{k_n}|)z,$$

where $y_2 = \theta_{t_0}(|a\rangle\langle e_{k_1}|)e_{k_2} = \theta_{t_0}(|a\rangle\langle a|)\theta_{t_0}(|a\rangle\langle e_{k_1}|)e_{k_2} \in \mathcal{H}_{t_0]}$. Then it is clear that by repeating this procedure we finally arrive at

$$\theta_{t+t_0}(|x_1\rangle\langle x_2|)\theta_t(|e_{k_1}\rangle\langle e_{k_1}|)\theta_{t-t_0}(|e_{k_2}\rangle\langle e_{k_2}|)\cdots\theta_{t-nt_0}(|e_{k_n}\rangle\langle e_{k_n}|)z$$

$$= \theta_t(|y_1\rangle\langle a|)\theta_{t-t_0}(|y_2\rangle\langle a|)\cdots\theta_{t-nt_0}(|y_{n+1}\rangle\langle a|)\theta_{t-nt_0}(|a\rangle\langle e_{k_n}|)z,$$

where $y_i \in \mathcal{H}_{t_0]}$ so that $|y_i\rangle\langle a| \in \mathcal{B}(\mathcal{H}_{t_0]})$ for $1 \leq i \leq n+1$. Moreover as $0 \leq t - nt_0 < t_0, \theta_{t-nt_0}(|a\rangle\langle e_{k_1}|)z = \theta_{t-nt_0}(|a\rangle\langle a|)\theta_{t-nt_0}(|a\rangle\langle e_{k_1}|)z \in \mathcal{H}_{t-nt_0]} \subset \mathcal{H}_{t_0]}$.

This completes the proof in view of (3.6).   ■

4 Domination for E_0-semigroups.

A quantum dynamical semigroup $\alpha = \{\alpha_t : t \geq 0\}$ is said to be dominated by a quantum dynamical semigroup $\tau = \{\tau_t : t \geq 0\}$, if $\tau_t - \alpha_t$ is completely positive for every t. We denote this by $\alpha \leq \tau$. Let $\mathcal{D}_\tau$ denote the set of all (not necessarily unital) quantum dynamical semigroups dominated by τ. $\mathcal{D}_\tau$ is a partially ordered set with partial order $\leq$. In this section we study this set for E_0-semigroups. We come across a situation where quite remarkably positivity implies complete positivity (Proposition 4.2).

Later we will see that $\mathcal{D}$ is a cocycle conjugacy invariant for E_0-semigroups. We can get information about the dilation by studying $\mathcal{D}_\tau$. For example, elementary semigroups in $\mathcal{D}_\tau$ correspond to units of the minimal dilation (up to scaling). In fact, precisely this is the fact used by Arveson to deduce that minimal dilations of unital quantum dynamical semigroups with bounded generators are cocycle conjugate to CCR flows [Ar5].

To begin with we show that quantum dynamical semigroups dominated by E_0-semigroups can be described through cocycles. The following result is more general than we are going to need.

Proposition 4.1: Let $\mathcal{A}, \mathcal{B}$ be unital C^*-algebras and α, β be linear maps from $\mathcal{A}$ to $\mathcal{B}$, where α is a unital $*$-homomorphism and β is completely positive. Suppose $\alpha - \beta$ is also completely positive. Then $\beta(X) = \alpha(X)\beta(I) = \beta(I)\alpha(X)$, for every $X \in \mathcal{A}$. (In particular $\beta(I)$ commutes with range of α.)

Proof: Let $\mathcal{B}$ be a unital subalgebra of $\mathcal{B}(\mathcal{H})$ for some Hilbert space $\mathcal{H}$.

Then Stinespring's theorem [St], applied to β provides us with a Hilbert space $\mathcal{K}$ with an isometry $V : \mathcal{H} \to \mathcal{K}$ and a representation $\pi : \mathcal{A} \to \mathcal{B}(\mathcal{K})$ such that $\beta(X) = V^*\pi(X)V$. Now for $X_1, X_2, \ldots, X_n, Y_1, Y_2, \ldots, Y_n$ in $\mathcal{A}$, we claim

$$|| \sum_{j=1}^{n} \beta(X_j)\alpha(Y_j)||^2 \leq || \sum_{j=1}^{n} \alpha(X_j)\alpha(Y_j)||^2.$$

Indeed considering the $n \times n$ block matrix $[\beta(X_i)^*\beta(X_j)]$, we have

$$
\begin{aligned}
[\beta(X_i)^*\beta(X_j)] &= [V^*\pi(X_i)^*VV^*\pi(X_j)V\ ] \\
&\leq [V^*\pi(X_i)^*\pi(X_j)V\ ] = [V^*\pi(X_i^*X_j)V\ ] = [\beta(X_i^*X_j)] \\
&\leq [\alpha(X_i^*X_j)] = [\alpha(X_i)^*\alpha(X_j)].
\end{aligned}
$$

Hence

$$
\begin{aligned}
(\sum_{i=1}^{n} \beta(X_i)\alpha(Y_i))^*(\sum_{i=1}^{n} \beta(X_i)\alpha(Y_i)) &= \sum_{i,j} \alpha(Y_i)^*\beta(X_i)^*\beta(X_j)\alpha(Y_j) \\
&\leq \sum_{i,j} \alpha(Y_i)^*\alpha(X_i)^*\alpha(X_j)\alpha(Y_j) \\
&= (\sum_{i=1}^{n} \alpha(X_i)\alpha(Y_i))^*(\sum_{i=1}^{n} \alpha(X_i)\alpha(Y_i)),
\end{aligned}
$$

proving our claim. Now take $n = 2, X_1 = I, X_2 = X, Y_1 = X, Y_2 = -I$, to obtain

$$||\beta(I)\alpha(X) - \beta(X)||^2 \leq ||\alpha(I)\alpha(X) - \alpha(X)||^2 = 0.$$

Therefore $\beta(X) = \beta(I)\alpha(X)$. On taking adjoints we have $\beta(X) = \alpha(X)\beta(I)$ $= \beta(I)\alpha(X)$ for all X in $\mathcal{A}$.   ■

Note that we needed just 2-positivity of $\alpha - \beta$. That itself forces the complete positivity for $\alpha - \beta$. Actually even 1-positivity is sufficient as the following result shows. The proof provided here arose out of a discussion with G.

Pisier. It may be compared with Rosenblum's proof of Fuglede-Putnam theorem (See Theorem 1.6.2 in [Pu]).

Proposition 4.2: Consider the set up of Proposition 4.1. The conclusions there hold even if $\alpha - \beta$ is assumed to be just positive.

Proof: Let $\beta(X) = V^*\pi(X)V, \quad X \in \mathcal{A}$, be the Stinespring representation of β as in the proof of Proposition 4.1. As $\alpha - \beta$ is positive, for any $X \in \mathcal{A}$, we have $V^*\gamma(X^*X)V \le \alpha(X^*X)$. Hence for any $z \in \mathcal{C}, H \in \mathcal{A}, u \in \mathcal{H}$

$$\|\gamma(e^{zH})Vu\|^2 \le \|\alpha(e^{zH})u\|^2.$$

Taking $u = \alpha(e^{-zH})v, v \in \mathcal{H}$, we have

$$|\langle v, \gamma(e^{zH})V\alpha(e^{-zH})v\rangle| \le \|v\|\|\gamma(e^{zH})Vu\| \le \|v\|^2.$$

Therefore the entire function $z \mapsto \langle v, \gamma(e^{zH})V\alpha(e^{-zH})v\rangle$ is bounded. Hence by Liouville's theorem it is a constant. So we get

$$\gamma(e^{zH})V\alpha(e^{-zH}) = \gamma(1)V\alpha(1) = \gamma(1)V,$$

or $\gamma(e^{zH})V = \gamma(1)V\alpha(e^{zH})$ for all $z \in \mathcal{C}, H \in \mathcal{A}$. This clearly implies $\gamma(X)V = \gamma(1)V\alpha(X)$ and hence $\beta(X) = V^*\gamma(X)V = V^*\gamma(1)V\alpha(X) = \beta(1)\alpha(X)$ for all $X \in \mathcal{A}$.   ∎

A much more direct proof of Proposition 4.2 can be given in the case we are presently concerned with namely when $\mathcal{A} = \mathcal{B} = \mathcal{B}(\mathcal{H})$ for a complex, separable Hilbert space $\mathcal{H}$, and $\alpha, \beta : \mathcal{B}(\mathcal{H}) \to \mathcal{B}(\mathcal{H})$ are normal. It is well-known that in such a case α, β have representations

$$\alpha(X) = \sum_n V_n X V_n^*, \ \beta(X) = \sum_k L_k X L_k^*, \ X \in \mathcal{B}(\mathcal{H}),$$

where $V_n, L_k \in \mathcal{B}(\mathcal{H}), \sum_n V_n V_n{}^*, \sum_k L_k L_k^*$ converge strongly (they could be finite sums) to $\alpha(I), \beta(I)$ respectively. Furthermore as α is a $*$ - endomorphism $\{V_n\}$'s can be chosen to be isometries with orthogonal range, i.e., $V_m{}^* V_n = \delta_{m,n}$ for $m, n \geq 0$. Fix $k, m \geq 0$. For $X \geq 0$ we have

$$L_k X L_k^* \leq \beta(X) \leq \sum_n V_n X V_n^*.$$

Multiplication by $V_m{}^*, V_m$ leads to $V_m{}^* L_k X L_k^* V_m \leq X$ for $X \geq 0$ in $\mathcal{B}(\mathcal{H})$. For $x \in \mathcal{H}$, we see that

$$V_m{}^* L_k |x\rangle\langle x| L_k^* V_m \leq |x\rangle\langle x|,$$

that is,

$$|V_m{}^* L_k x\rangle\langle V_m^* L_k x| \leq |x\rangle\langle x|.$$

But then, as rank-one operators are extremal in the cone of positive operators, we obtain

$$V_m{}^* L_k x = c_{m,k}(x) x.$$

for some $c_{m,k}(x) \in \mathbb{C}$. However, as this happens for every vector $x \in \mathcal{H}$, it is not difficult to see that $c_{m,k}(x) \equiv c_{m,k}$, for some $c_{m,k} \in \mathbb{C}$. In other words, for every m, k $V_m{}^* L_k = c_{m,k} I$. Hence

$$
\begin{aligned}
V_m X V_m{}^* \cdot \beta(I) &= V_m X V_m{}^* (\sum_k L_k L_k^*) = \sum_k V_m X c_{m,k} L_k^* \\
&= \sum_k V_m V_m{}^* L_k X L_k^* \\
&= V_m V_m{}^* \beta(X).
\end{aligned}
$$

Adding this over m we have $\alpha(X)\beta(I) = \beta(X)$. These arguments also show that if the map $X \mapsto L X L^*$ is dominated by a $*$-endomorphism $\alpha(X) = \sum_n V_n X V_n{}^*$ then $L = \sum_n c_n V_n$, and consequently, $L^* L$ is a scalar.

Now let θ be an E_0-semigroup of $\mathcal{B}(\mathcal{H})$. Recall the definition of local cocycles (Definition 2.3). It is clear that if G is a positive local cocycle of θ, then $\psi_t(\cdot) = G_t\theta(\cdot)$ is a quantum dynamical semigroup.

Theorem 4.3: Let θ be an E_0-semigroup of $\mathcal{B}(\mathcal{H})$, and let ψ be a quantum dynamical semigroup of $\mathcal{B}(\mathcal{H})$. Then the following are equivalent:

(i) θ dominates ψ;

(ii) $\theta_t - \psi_t$ is positive for every $t \geq 0$;

(iii) $\psi_t(Z) = G_t\theta_t(Z), \forall Z \in \mathcal{B}(\mathcal{H})$, for some positive, contractive, local cocycle of θ;

(iv) ψ is *absorbing* for θ, that is, $\psi_t(Z)\theta_t(W) = \psi_t(ZW), \forall Z, W \in \mathcal{B}(\mathcal{H}), t \geq 0$.

Proof: (i) implies (ii) is trivial. Assuming (ii), take $G_t = \psi_t(I)$, then $0 \leq G_t \leq I$, and from Proposition 4.2 we have $G_t \in \theta_t(\mathcal{B}(\mathcal{H}))'$, and $\psi_t(Z) = G_t\theta_t(Z)$, for $Z \in \mathcal{B}(\mathcal{H}), t \geq 0$. Now $G_{s+t} = \psi_{s+t}(I) = G_s\theta_s(\psi_t(I)) = G_s\theta_s(G_t)$, for $s, t \geq 0$. Further $t \mapsto G_t$ is strongly continuous as $G_t = \psi_t(1)$ (see Proposition A.1). This proves (ii) implies (iii). Clearly (iii) is equivalent to (iv). Finally to prove that (iii) implies (i), observe $\theta_t(Z) = G_t\theta_t(Z) + (I - G_t)\theta_t(Z) = \psi_t(Z) + (I - G_t)^{\frac{1}{2}}\theta_t(Z)(I - G_t)^{\frac{1}{2}}$. Hence $\psi \leq \theta$. $\blacksquare$

Though we have not stated Theorem 4.3 for discrete quantum dynamical semigroups, the analogus result holds. For instance in Example 2.1, quantum dynamical semigroups dominated by $\{\theta^n : n \geq 0\}$ are of the form

$\{\psi^n : n \geq 0\}$, where

$$\psi(X_1 \otimes X_2 \otimes \cdots) = G \otimes X_1 \otimes X_2 \otimes \cdots,$$

for some positive contraction G.

Proposition 4.4: The partially ordered set $\mathcal{D}$ of dominated quantum dynamical semigroups is a cocycle conjugacy invariant for E_0-semigroups.

Proof: Let θ, θ' be E_0-semigroups of $\mathcal{B}(\mathcal{H})$, such that

$$\theta'_t(Z) = U_t \theta_t(Z) U_t^*$$

for a unitary cocycle $\{U_t\}$ of θ. Then it is clear that $K = \{K_t : t \geq 0\}$ is a positive, contractive, local cocycle of θ' iff $K_t = U_t G_t U_t^*$ for a cocycle $G = \{G_t : t \geq 0\}$ of θ.   ∎

We also remark that for an E_0-semigroup θ, with local cocycle $\{G_t\}$, $\psi_t(Z) = \theta_t(Z)G_t$, $t \geq 0$ is an e_0-semigroup iff $\{G_t\}$ is a projection cocycle. For example, the induced semigroup $\tilde{\tau}$ of Section 3 is given by

$$\tilde{\tau}_t(Z) = \theta_t(Z)E_t, \quad E_t = \tilde{\tau}_t(1) = W_t(1_{\mathcal{H}} \otimes \hat{\mathcal{P}}_t)W_t^* \tag{4.1}$$

and E_t is the associated projection cocyle. Of course, the partial ordered set $\mathcal{E}$ of dominated e_0-semigroups is also a cocycle conjugacy invariant. Powers [Po3] has shown that this set is actually a complete lattice, after adding a 'zero' element which is dominated by all e_0-semigroups.

5   Compression under domination

Let $\mathcal{H}$ be a complex separable Hilbert space, and let θ be an E_0-semigroup of $\mathcal{B}(\mathcal{H})$. Let $\mathcal{H}_0$ be a subspace of $\mathcal{H}$ and suppose that the projection P onto $\mathcal{H}_0$ is increasing (recall the definition from Section 3) for θ. Let τ be the unital quantum dynamical semigroup got by compressing θ by P. Throughout this section we will assume this set up.

At first we will show that compression by projection P maps $\mathcal{D}_\theta$ surjectively to $\mathcal{D}_\tau$. (A priori it is not even clear that compressions of elements in $\mathcal{D}_\theta$ are quantum dynamical semigroups.) The main theorem in this section is that this compression map is injective if and only if θ is the minimal dilation of τ. This gives us a completely algebraic characterization of minimality. The usefulness of this result will be obvious in subsequent sections, in fact this will be the main tool in deciding minimality of many dilations.

Proposition 5.1: Suppose ψ is a quantum dynamical semigroup dominated by θ then its compression by P is a quantum dynamical semigroup dominated by τ.

Proof: Let $\alpha = \{\alpha_t; t \geq 0\}$ be the compression of ψ with respect to P, that is,

$$\alpha_t(X) = P\psi_t(X)P, \quad X \in \mathcal{B}(\mathcal{H}_0) = P\mathcal{B}(\mathcal{H})P$$

for $t \geq 0$. It is clear that α_t is a contractive, completely positive map for every t, and $\alpha_0(X) = P\psi_0(X)P = X$, for $X \in \mathcal{B}(\mathcal{H}_0)$. Now by Theorem 4.3, on taking $G_t = \psi_t(I)$, $G = \{G_t, t \geq 0\}$ is a local, positive, contractive

33

cocycle of θ such that $\psi_t(Z) = G_t \theta_t(Z)$ for $Z \in \mathcal{B}(\mathcal{H})$. So for $s, t \geq 0$ and $X \in \mathcal{B}(\mathcal{H}_0)$,

$$
\begin{aligned}
\alpha_s(\alpha_t(X)) &= \alpha_s(PG_t\theta_t(X)P) = PG_s\theta_s(PG_t\theta_t(X)P)P \\
&= PG_s\theta_s(P)\theta_s(G_t)\theta_{s+t}(X)\theta_s(P)P \\
&= P\theta_s(P)G_{S+t}\theta_{s+t}(X)\theta_s(P)P \\
&= PG_{s+t}\theta_{s+t}(X)P = P\psi_{s+t}(X)P \\
&= \alpha_{s+t}(X).
\end{aligned}
$$

Normality and weak operator continuity of α are obvious as ψ has these properties. As θ dominates ψ it is also clear that τ dominates α. ∎

Fixing the increasing projection P we denote the associated compression map by $\mu = \mu^P$. What we have shown is that μ maps $\mathcal{D}_\theta$ (the partially ordered set of quantum dynamical semigroups dominated by θ) to $\mathcal{D}_\tau$ where $\mu(\theta) = \tau$. It is clear that μ respects the partial order. Now we are going to show that this map is always surjective. It is injective (hence an isomorphism) if and only if θ is the minimal dilation of τ.

Let α be a quantum dynamical semigroup dominated by τ. We want to obtain a quantum dynamical semigroup ψ dominated by θ such that its compression is α. As in Section 2, let $\hat{\tau}$ acting on $\mathcal{B}(\hat{\mathcal{H}})$ be the minimal dilation of τ and let $\tilde{\tau}$ be the induced semigroup. Note that $\tilde{\tau}$ is dominated by θ.

The following lemma comes from an old argument in [Bh1]. However, for readers' convenience we sketch the proof here.

Lemma 5.2 : Fix $t \geq 0$. Then there exists a unique positive contraction $C_t \in \mathcal{B}(\hat{\mathcal{H}}_{t]})$ such that,

$$\langle \theta(\underline{r}, \underline{Y})u, C_t\theta(\underline{r}, \underline{Z})v\rangle = \langle u, \alpha_{r_n}(Y_n{}^* \cdots \alpha_{r_2-r_3}(Y_2{}^*\alpha_{r_1-r_2}(Y_1{}^*Z_1)Z_2) \cdots Z_n)v\rangle$$

for $(\underline{r}, \underline{Y}, u), (\underline{r}, \underline{Z}, v) \in \mathcal{N}_t$. Moreover C_t commutes with $\theta_t(X)$ for every $X \in \mathcal{B}(\mathcal{H}_0)$.

Proof : Let $\alpha' = \{\alpha'_t : t \geq 0\}$ acting on $\mathcal{B}(\mathcal{H}')$ be the unique minimal dilation of quantum dynamical semigroup α. (Note that α may not be unital, but this is no problem for existence of minimal dilation [Bh3]). Define a map $D_t : \hat{\mathcal{H}}_{t]} \to \mathcal{H}'$ by

$$D_t\theta(\underline{r}, \underline{Y})u = \alpha'(\underline{r}, \underline{Y})u \qquad \text{for } (\underline{r}, \underline{Y}, u) \in \mathcal{N}_t,$$

and extending linearly. To verify the well-definedness of D_t we need to estimate

$$\sum_{i,k=1}^{p} \bar{c}_i c_k \langle \alpha'(\underline{r}^{(i)}, \underline{Y}^{(i)})u^{(i)}, \alpha'(\underline{r}^{(k)}, \underline{Y}^{(k)})u^{(k)}\rangle$$

for arbitrary $c_i \in \mathbb{C}$, $(\underline{r}^{(i)}, \underline{Y}^{(i)}, u^{(i)}) \in \mathcal{N}_t$, $1 \leq i \leq p$. A basic property of minimal dilation is that ([Bh1], [BP1]) for $(\underline{s}, \underline{X}, u) \in \mathcal{N}_t$, $r \leq t$

$$\alpha'_{s_1}(X_1) \cdots \alpha'_{s_n}(X_n)u = \alpha'_{s_1}(X_1) \cdots \alpha'_{s_{i-1}}(X_{i-1})\alpha'_r(I)\alpha'_{s_i}(X_i) \cdots \alpha'_{s_n}(X_n)u$$

if $s_{i-1} \geq r \geq s_i$. In other words we can always insert $\alpha_r(I)$ at appropriate place (so as to keep the descending order of time points) for any time point $r \leq s_1 = t$ without changing the vector. It follows that without loss of generality we can take $\underline{r}^{(1)} = \underline{r}^{(2)} = \cdots = \underline{r}^{(p)} = \underline{r} = (r_1, \cdots, r_n)$, in the inner-product we want to estimate. If $\underline{Y}^{(i)} = (Y_{1i}, \ldots, Y_{ni})$, $1 \leq i \leq p$

consider the operator $A = ((A_{ik}))_{1 \le i,k \le p}$, on $\mathcal{H}_0 \oplus \mathcal{H}_0 \cdots \oplus \mathcal{H}_0$ (p copies) defined by $A_{ik} =$

$$\tau_{r_n}(Y_{ni}^{\;*}\tau_{r_{n-1}-r_n}(Y_{(n-1)i}^{\;*} \cdots \tau_{r_2-r_1}(Y_{2i}^{\;*}\tau_{r_1-r_2}(Y_{1i}^{\;*}Y_{1k})Y_{2k}) \cdots Y_{(n-1)k})Y_{nk}).$$

Let A' be the operator defined in similar fashion using α instead of τ. Then the assumption that τ dominates α coupled with a simple induction argument shows that $A' \le A$. Now

$$
\begin{aligned}
\|D_t \sum_{i=1}^{p} c_i \theta(\underline{r}, \underline{Y}^{\;(i)}) u^{(i)}\|^2
&= \sum_{i,k=1}^{p} \bar{c}_i c_k \langle \alpha'(\underline{r}, \underline{Y}^{\;(i)}) u^{(i)}, \alpha'(\underline{r}, \underline{Y}^{\;(k)},) u^{(k)} \rangle \\
&= \langle (\oplus_i c_i u_i,), A'(\oplus_k c_k u_k) \rangle \\
&\le \langle (\oplus_i c_i u_i,), A(\oplus_k c_k u_k) \rangle \\
&= \| \sum_{i=1}^{p} c_i \theta(\underline{r}, \underline{Y}^{\;(i)}) u^{(i)} \|^2.
\end{aligned}
$$

This shows that D_t extends to a linear contraction on $\hat{\mathcal{H}}_{t]}$. Now take $C_t = D_t^* D_t$. It is clear that C_t satisfies the requirement. Uniqueness follows as $\overline{\text{span}}\{\theta(\underline{r}, \underline{Y})u : (\underline{r}, \underline{Y}, u) \in \mathcal{N}_t\} = \hat{\mathcal{H}}_{t]}$. Moreover for $X \in \mathcal{B}(\mathcal{H}_0)$, and $(\underline{r}, \underline{Y}, u), (\underline{r}, \underline{Z}, v) \in \mathcal{N}_t$,

$$
\begin{aligned}
&\langle \theta(\underline{r}, \underline{Y})u, C_t \theta_t(X) \theta(\underline{r}, \underline{Z})v \rangle \\
&= \langle u, \alpha_{r_n}(Y_n^{\;*} \cdots \alpha_{r_2-r_3}(Y_2^{\;*}\alpha_{r_1-r_2}(Y_1^{\;*} X Z_1)Z_2) \cdots Z_n)v \rangle \\
&= \langle \theta_t(X^*) \theta(\underline{r}, \underline{Y})u, C_t \theta(\underline{r}, \underline{Z})v \rangle \\
&= \langle \theta(\underline{r}, \underline{Y})u, \theta_t(X) C_t \theta(\underline{r}, \underline{Z})v \rangle.
\end{aligned}
$$

Therefore C_t commutes with $\theta_t(X)$, for $X \in \mathcal{B}(\mathcal{H}_0)$.   ∎

Note that for $X \in \mathcal{B}(\mathcal{H}_0), \theta_t(X)$ on $\hat{\mathcal{H}}_{t]}$ is just $\hat{\tau}_t(X)$, (see 3.1) so C_t commutes with $\hat{\tau}_t(X)$. Furthermore from Proposition 3.1, $\hat{\tau}_t(X) = W_t(X \otimes$

$1_{\hat{\mathcal{P}}_t})W_t{}^*, W_t(\mathcal{H}_0 \otimes \hat{\mathcal{P}}_t) = \hat{\mathcal{H}}_{t]}$, we obtain

$$C_t = W_t(1_{\mathcal{H}_0} \otimes \bar{C}_t)W_t{}^* \quad t \geq 0$$

for some $\bar{C}_t \in \mathcal{B}(\hat{\mathcal{P}}_t)$. Define $\hat{G} = \{\hat{G}_t : t \geq 0\}, \tilde{G} = \{\tilde{G}_t : t \geq 0\}$ by setting

$$\hat{G}_t = W_t(1_{\hat{\mathcal{H}}} \otimes \bar{C}_t)W_t{}^* \qquad \text{(acting on } \hat{\mathcal{H}})$$

$$\tilde{G}_t = W_t(1_{\mathcal{H}} \otimes \bar{C}_t)W_t{}^* \qquad \text{(acting on } \mathcal{H})$$

for $t \geq 0$.

Theorem 5.3: Define $\hat{\alpha} = \{\hat{\alpha}_t : t \geq 0\}, \tilde{\alpha} = \{\tilde{\alpha}_t : t \geq 0\}$ by

$$\hat{\alpha}_t(Y) = \hat{G}_t \hat{\tau}_t(Y) \quad \text{for} \ Y \in \mathcal{B}(\hat{\mathcal{H}})$$

$$\tilde{\alpha}_t(Z) = \tilde{G}_t \tilde{\tau}_t(Z) \quad \text{for} \ Z \in \mathcal{B}(\mathcal{H})$$

Then $\hat{\alpha}, \tilde{\alpha}$ are unique quantum dynamical semigroups dominated by $\hat{\tau}, \tilde{\tau}$ respectively such that their compression to $\mathcal{B}(\mathcal{H}_0)$ is α. Moreover $\tilde{\alpha}$ is dominated by θ.

Proof : In view of Theorem 4.3, we need to show that $\hat{G}, \tilde{G}$ are positive, contractive, local cocycles of $\hat{\tau}, \tilde{\tau}$. Positivity, contractivity are clear as $\bar{C}_t$ has these properties. Now let us consider $\hat{G}$. As $\hat{\tau}_t(Y) = W_t(Y \otimes 1_{\hat{\mathcal{P}}_t})W_t^*$, for $Y \in \mathcal{B}(\hat{\mathcal{H}})$, $\hat{\tau}_t(Y)$ commutes with $\hat{G}_t$ for every t. But then as $\hat{\tau}$ is a semigroup $\hat{G}_t$ commutes with $\hat{\tau}_s(Y)$ for all $s \geq t$. Also note that $1_{\hat{\mathcal{H}}_{t]}} = W_t(1_{\mathcal{H}_0} \otimes 1_{\hat{\mathcal{P}}_t})W_t^*$. Hence $1_{\hat{\mathcal{H}}_{t]}} \hat{G}_t 1_{\hat{\mathcal{H}}_{t]}} = C_t$. Now consider $(\underline{r}, \underline{Y}, u), (\underline{r}, \underline{Z}, v) \in \mathcal{N}$. Without loss of generality we assume $t = r_k$ for some k. Then by moment formula (3.2),

$$\langle \hat{\tau}(\underline{r}, \underline{Y})u, \hat{G}_t \hat{\tau}(\underline{r}, \underline{Z})v \rangle$$

$$= \langle \hat{\tau}_{r_{k+1}}(Y_{k+1}) \cdots \hat{\tau}_{r_n}(Y_n)u,$$

$$\hat{G}_t[\hat{\tau}_{r_k}(Y_k{}^*)\cdots\hat{\tau}_{r_1}(Y_1{}^*)\hat{\tau}_{r_1}(Z_1)\cdots\hat{\tau}_{r_k}(Z_k)]\hat{\tau}_{r_{k+1}}(Z_{k+1})\cdots\hat{\tau}_{r_n}(Z_n)v\rangle$$

$$= \langle\hat{\tau}_{r_{k+1}}(Y_{k+1})\cdots\hat{\tau}_{r_n}(Y_n)u,$$

$$C_t\hat{\tau}_{r_k}(Y_k{}^*\cdots\tau_{r_2-r_3}(Y_2^*\tau_{r_1-r_2}(Y_1{}^*Z_1)Z_2)\cdots Z_k)\hat{\tau}_{r_{k+1}}(Z_{k+1})\cdots\hat{\tau}_{r_n}(Z_n)v\rangle$$

Now by Lemma 5.2 (definition of C_t)

$$\langle\hat{\tau}(\underline{r},\underline{Y})u,\hat{G}_t\hat{\tau}(\underline{r},\underline{Z})v\rangle$$

$$= \langle u,\alpha_{r_n}(Y_n{}^*\cdots Y_{k+1}^*\alpha_{r_k-r_{k+1}}(Y_k{}^*\tau_{r_{k-1}-r_k}(Y_{k-1}^*\cdots$$

$$\tau_{r_2-r_3}(Y_2{}^*\tau_{r_1-r_2}(Y_1{}^*Z_1)Z_2)\cdots Z_{k-1})Z_k)Z_{k+1}\cdots Z_n)v\rangle(5.1)$$

Then to arrive at cocycle property consider $(\underline{r},\underline{Y},u),(\underline{r},\underline{Z},v)\in\mathcal{N}$ and $s,t\geq$ 0. This time without loss of generality assume $s+t=r_k$ for some k, $s=r_m$ for some m. We have

$$\langle\hat{\tau}(\underline{r},\underline{Y})u,\hat{G}_s\hat{\tau}_s(\hat{G}_t)\hat{\tau}(\underline{r},\underline{Z})v\rangle$$

$$= \langle\hat{\tau}_{r_{m+1}}(Y_{m+1})\cdots\hat{\tau}_{r_n}(Y_n)u,\hat{G}_s\hat{\tau}_s(M)\hat{\tau}_{r_{m+1}}(Z_{m+1})\cdots\hat{\tau}_{r_n}(Z_n)v\rangle,$$

where

$$M = Y_m{}^*(\hat{\tau}_{r_{m-1}-s}(Y_{m-1}^*)\cdots\hat{\tau}_{r_1-s}(Y_1{}^*)\hat{G}_t\hat{\tau}_{r_1-s}(Z_1)\cdots\hat{\tau}_{r_{m-1}-s}(Z_{m-1})Z_m.$$

But then by computation of moments for $\hat{G}_t$ (5.1) (note that now $r_k-s=t$)

$$M = Y_m{}^*\alpha_{r_{m-1}-s}(Y_{m-1}^*\cdots\alpha_{r_k-r_{k+1}}(Y_k{}^*\tau_{r_{k-1}-r_k}(Y_{k-1}^*$$

$$\cdots\tau_{r_2-r_3}(Y_2{}^*\tau_{r_1-r_2}(Y_1{}^*Z_1)Z_2)\cdots Z_{k+1})Z_k)\cdots Z_{m-1})Z_m,$$

and

$$\langle\hat{\tau}_{r_{m+1}}(Y_{m+1})\cdots\hat{\tau}_{r_n}(Y_n)u,\hat{G}_s\hat{\tau}_s(M)\hat{\tau}_{r_{m+1}}(Z_{m+1})\cdots\hat{\tau}_{r_n}(Z_n)v\rangle$$

$$= \langle u,\alpha_{r_n}(Y_n{}^*\cdots\alpha_{r_{m+1}-r_{m+2}}(Y_{m+1}\alpha_{r_m-r_{m+1}}(M)Z_{m+1})\cdots Z_n)v\rangle.$$

Substituting the formula for M we get

$$\langle \hat{\tau}(\underline{r}, \underline{Y})u, \hat{G}_s \hat{\tau}_s(\hat{G}_t) \hat{\tau}(\underline{r}, \underline{Z})v \rangle = \langle \hat{\tau}(\underline{r}, \underline{Y})u, \hat{G}_{s+t} \hat{\tau}(\underline{r}, \underline{Z})v \rangle.$$

By totality of vectors of the form $\{\hat{\tau}(\underline{r}, \underline{Y})u : (\underline{r}, \underline{Y}, u) \in \mathcal{N}\}$ in $\hat{\mathcal{H}}$ we obtain $\hat{G}_s \hat{\tau}_s(\hat{G}_t) = \hat{G}_{s+t}$. This shows that $\hat{G}$ is a cocycle for $\hat{\tau}$. (Continuity follows from Appendix A, Proposition A.5). Futhermore,

$$
\begin{aligned}
\hat{G}_s \hat{\tau}_s(\hat{G}_t) &= W_s(1_{\hat{\mathcal{H}}} \otimes \bar{C}_s)W_s{}^* W_s(\hat{G}_t \otimes 1_{\hat{\mathcal{P}}_s})W_s{}^* \\
&= W_s(\hat{G}_t \otimes \bar{C}_s)W_s{}^* \\
&= W_s(W_t(1_{\hat{\mathcal{H}}} \otimes \bar{C}_t)W_t{}^* \otimes \bar{C}_s)W_s{}^* \\
&= W_s(W_t \otimes 1_{\mathcal{P}_s})(1_{\hat{\mathcal{H}}} \otimes \bar{C}_t \otimes \bar{C}_s)(W_t{}^* \otimes 1_{\mathcal{P}_s})W_s{}^*.
\end{aligned}
$$

But recall from Proposition 3.2(i) that $W_s(W_t \otimes 1_{\mathcal{P}_s}) = W_{s+t}(1_{\mathcal{H}} \otimes U_{t,s})$ on $\mathcal{H} \otimes \mathcal{P}_t \otimes \mathcal{P}_s$. Hence

$$
\begin{aligned}
W_{s+t}(1_{\hat{\mathcal{H}}} \otimes \bar{C}_{s+t})W_{s+t}^* &= \hat{G}_{s+t} = \hat{G}_s \hat{\tau}_s(\hat{G}_t) \\
&= W_{s+t}(1_{\hat{\mathcal{H}}} \otimes U_{t,s}(\bar{C}_t \otimes \bar{C}_s)U_{t,s}^*)W_{s+t}^*,
\end{aligned}
$$

proving

$$U_{t,s}(\bar{C}_t \otimes \bar{C}_s)U_{t,s}^* = \bar{C}_{s+t}.$$

Now

$$
\begin{aligned}
\tilde{G}_s \tilde{\tau}_s(\tilde{G}_t) &= W_s(1_{\mathcal{H}} \otimes \bar{C}_s)W_s{}^* W_s(\tilde{G}_t \otimes 1_{\hat{\mathcal{P}}_s})W_s{}^* \\
&= W_s(\tilde{G}_t \otimes \bar{C}_s)W_s{}^* \\
&= W_s(W_t \otimes 1_{\mathcal{P}_s})(1_{\mathcal{H}} \otimes \bar{C}_t \otimes \bar{C}_s)(W_t{}^* \otimes 1_{\mathcal{P}_s})W_t \\
&= W_{s+t}(1_{\mathcal{H}} \otimes U_{t,s})(1_{\mathcal{H}} \otimes \bar{C}_t \otimes \bar{C}_s)(1_{\mathcal{H}} \otimes U_{t,s}^*)W_{s+t}^* \\
&= W_{s+t}(1_{\mathcal{H}} \otimes \bar{C}_{s+t})W_{s+t}^* \\
&= \tilde{G}_{s+t}.
\end{aligned}
$$

Hence $\tilde{G}$ s a cocycle for $\tilde{\tau}$. Finally for $X_1, X_2 \in \mathcal{B}(\mathcal{H}_0), u_1, u_2 \in \mathcal{H}_0, t \geq 0$

$$
\begin{aligned}
\langle \tilde{\tau}_t(X_1)u_1, \tilde{G}_t \tilde{\tau}_t(X_2)u_2 \rangle &= \langle u_1, W_t(X_1^* X_2 \otimes \bar{C}_t)W_t^* u_2 \rangle \\
&= \langle \hat{\tau}_t(X_1)u_1, \hat{G}_t \hat{\tau}_t(X_2)u_2 \rangle \\
&= \langle \hat{\tau}_t(X_1)u_1, C_t \hat{\tau}_t(X_2)u_2 \rangle \\
&= \langle \theta_t(X_1)u_1, C_t \theta_t(X_2)u_2 \rangle \\
&= \langle u_1, \alpha_t(X_1^* X_2)u_2 \rangle.
\end{aligned}
$$

Therefore the compression of $\tilde{\alpha}, \hat{\alpha}$ to $\mathcal{B}(\mathcal{H}_0)$ is α, (Take $X_1 = P, X_2 = X \in \mathcal{B}(\mathcal{H}_0)$). As $\tilde{\tau} \leq \theta$, and $\tilde{\alpha} \leq \tilde{\tau}$ we have $\tilde{\alpha} \leq \theta$. To prove uniqueness of $\hat{\alpha}$, or equivalently that of $\hat{G}$, observe that we clearly need $\langle \theta_t(Y)u, \hat{G}_t \theta_t(Z)v \rangle = \langle u, \hat{\alpha}_t(Y^* Z)v \rangle = \langle u, \alpha_t(Y^* Z)v \rangle$, for $u, v \in \mathcal{H}_0, Y, Z \in \mathcal{B}(\mathcal{H}_0)$. Now local cocycle property of $\hat{G}_t$ and an easy induction argument shows that $\hat{G}_t$ must satisfy (5.1). So uniqueness follows from totality of vectors of the form $\{\hat{\tau}(\underline{r}, \underline{Y})u : (\underline{r}, \underline{Y}, u) \in \mathcal{N}\}$ in $\hat{\mathcal{H}}$. As for $\tilde{G}_t$, it is local implies that it is of the form $W_t(1_{\mathcal{H}} \otimes \bar{C}_t)W_t^*$ for some operator $\bar{C}_t$ on $\hat{\mathcal{P}}_t$. But then $\bar{C}_t$ is already determined because, the compression of $\tilde{G}_t$ to $\hat{\mathcal{H}}$ must be $\hat{G}_t$. $\blacksquare$

The content of Theorem 5.3, can be shown pictorially as:

$$
\begin{array}{ccccc}
\tilde{\alpha} & \leq & \tilde{\tau} & \leq & \theta \\
\downarrow & & \downarrow & \nearrow & \\
\hat{\alpha} & \leq & \hat{\tau} & & \\
\downarrow & & \downarrow & & \\
\alpha & \leq & \tau & &
\end{array}
$$

where as before $\leq$ denotes domination and arrows indicate compressions by appropriate projections. (Here α, τ act on $\mathcal{B}(\mathcal{H}_0), \hat{\alpha}, \hat{\tau}$ act on $\mathcal{B}(\hat{\mathcal{H}})$ and $\tilde{\alpha}, \tilde{\tau}, \theta$ act on $\mathcal{B}(\mathcal{H})$). As such a construction is possible for every α in $\mathcal{D}_\tau$, we have shown that the compression map μ maps $\mathcal{D}_\theta$ surjectively to $\mathcal{D}_\tau$. In fact, it maps $\mathcal{D}_{\tilde{\tau}}$ surjectively to $\mathcal{D}_\tau$. Furthermore, it is injective on $\mathcal{D}_{\tilde{\tau}}$, and the map factors through $\mathcal{D}_{\hat{\tau}}$. Hence $\mathcal{D}_{\tilde{\tau}}, \mathcal{D}_{\hat{\tau}}$ and $\mathcal{D}_\tau$ are isomorphic. Note that $\mathcal{D}_\theta, \mathcal{D}_{\tilde{\tau}}, \mathcal{D}_{\hat{\tau}}$ are described by cocycles unlike $\mathcal{D}_\tau$.

If $\tilde{\tau} \neq \theta$, then clearly injectivity fails to hold as both $\tilde{\tau}$ and θ are mapped to τ. This also shows that to verify minimality of dilation θ it is enough to check injectivity of the compression map on e_0-semigroups dominated by θ, i.e., it suffices to look at projection cocycles. Sometimes it is enough to look at even smaller classes (See Corollary 7.7).

Theorem 5.4: Let $\{E_t : t \geq 0\}$ be the projection cocycle of θ such that $\tilde{\tau}_t(\cdot) = \theta_t(\cdot)E_t$. If ψ is a semigroup in $\mathcal{D}_\theta$, then there exists a unique semigroup η in $\mathcal{D}_{\tilde{\tau}}$, having same compression (to $\mathcal{B}(\mathcal{H}_0)$) as ψ. If $\{G_t : t \geq 0\}$ is the cocycle of ψ then $\{E_t G_t E_t : t \geq 0\}$ is the cocycle of η.

Proof : We have $E_t = \tilde{\tau}_t(1) = W_t(1_{\mathcal{H}} \otimes 1_{\hat{\mathcal{P}}_t})W_t^*$, $t \geq 0$. As contractive local cocycles form a semigroup (see remarks on Definition 2.3), we know that $\{E_t G_t E_t : t \geq 0\}$ is a positive, contractive, local cocycle. Let η be the semigroup with this cocycle. As G_t commutes with $\theta_t(\mathcal{B}(\mathcal{H})), G_t = W_t(1_{\mathcal{H}} \otimes C_t)W_t^*$, for some $C_t \in \mathcal{B}(\mathcal{P}_t)$. Therefore $\eta_t(Z) = W_t(Z \otimes 1_{\hat{\mathcal{P}}_t} C_t 1_{\hat{\mathcal{P}}_t})W_t^*$, for $Z \in \mathcal{B}(\mathcal{H})$. Now as $1_{\hat{\mathcal{H}}} = W_t(1_{\hat{\mathcal{H}}} \otimes 1_{\hat{\mathcal{P}}_t})W_t^*$, compressing η to $\mathcal{B}(\hat{\mathcal{H}})$ we obtain

$$1_{\hat{\mathcal{H}}}\eta_t(Y)1_{\hat{\mathcal{H}}} = W_t(Y \otimes 1_{\hat{\mathcal{P}}_t} C_t 1_{\hat{\mathcal{P}}_t})W_t^* = 1_{\hat{\mathcal{H}}}\psi_t(Y)1_{\hat{\mathcal{H}}}, \quad \text{for } Y \in \mathcal{B}(\hat{\mathcal{H}}).$$

This shows that compression of ψ, η to $\mathcal{B}(\hat{\mathcal{H}})$ and hence to $\mathcal{B}(\mathcal{H}_0)$ are same. The uniqueness is clear as the compression map is injective on $\mathcal{D}_{\tilde{\tau}}$. ∎

Corollary 5.5: The cocycle $\{E_t : t \geq 0\}$ with $\tilde{\tau}_t(\cdot) = \theta_t(\cdot)E_t$, is the smallest cocycle of θ such that the compression of $\theta_t(\cdot)E_t$ to $\mathcal{B}(\mathcal{H}_0)$ is τ.

Proof : If $\{F_t : t \geq 0\}$ is another cocycle such that the compression of $\theta_t(\cdot)F_t$ to $\mathcal{B}(\mathcal{H}_0)$ is τ, then $\{E_t F_t E_t : t \geq 0\}$ also has the same property. By uniqueness in Theorem 5.4, $E_t F_t E_t = E_t$, for all t. As F_t is contractive, positive, and E_t is a projection this implies $F_t \geq E_t$ for every t. ∎

6   Units

Units play an important role in classification scheme of E_0-semigroups as developed by Powers and Arveson. The notion of units can be extended to quantum dynamical semigroups [Ar4]. Then we can study the behaviour of units under dilation. This has been carried out by Arveson [Ar5] for semigroups with bounded generators. He shows that units can always be lifted to a unit of minimal dilation. We give an explicit formula (6.2) for such a lifting. It is possible to give several alternative descriptions of units. One of them, connects units to extensions of quantum dynamical semigroups on a larger algebra. This helps us to solve a problem of Davies [Da4] regarding extensions of quantum dynamical semigroups.

Definition 6.1: Let $\tau = \{\tau_t : t \geq 0\}$ be a quantum dynamical semigroup of $\mathcal{B}(\mathcal{H}_0)$. Then a strongly continuous one parameter semigroup $A = \{A_t : t \geq 0\}$ of bounded operators on $\mathcal{H}_0$, is said to be a unit of τ if there exists a positive real number c such that τ dominates the elementary quantum dynamical semigroup $e^{-ct}\alpha^A$, $\alpha_t^A(X) = A_t X A_t^*$, $X \in \mathcal{B}(\mathcal{H}_0)$. The unit A is said to be normalized if c can be taken to be zero.

We denote the set of units of τ by $\mathcal{U}(\tau)$, and the set of normalized units by $\mathcal{U}_0(\tau)$. It is to be noted that if $\{A_t\} \in \mathcal{U}(\tau)$, then $\{e^{-qt}A_t\} \in \mathcal{U}_0(\tau)$ for suitable q. So there is no harm in restricting our attention to $\mathcal{U}_0(\tau)$. It may also be observed that as τ is assumed to be contractive A_t is contractive for every t if $\{A_t\} \in \mathcal{U}_0(\tau)$.

Units of E_0-semigroups can be described in different ways. Let $\theta = \{\theta_t : t \geq$

$0\}$ be an E_0-semigroup of $\mathcal{B}(\mathcal{H})$. Suppose $V = \{V_t : t \geq 0\}$ is a normalized unit of θ. As θ dominates the semigroup $\psi_t(\cdot) = V_t(\cdot)V_t{}^*$, from Theorem 4.3, we obtain that $\psi_t(1) = V_t V_t{}^*$ is a cocycle of θ and that

$$\theta_t(Z)V_t V_t{}^* = V_t Z V_t{}^* = V_t V_t{}^* \theta_t(Z), \quad \forall Z \in \mathcal{B}(\mathcal{H}).$$

But from remarks made just after Proposition 4.2, we know that $V_t^* V_t$ is a scalar for every t. As V is a contractive semigroup, we obtain $V_t{}^* V_t = e^{-ct}$, for some $c \geq 0$. This also yields,

$$\theta_t(Z)V_t = V_t Z, \quad \forall t \geq 0, \quad Z \in \mathcal{B}(\mathcal{H}_0). \tag{6.1}$$

Then it is clear that a unit V of an E_0-semigroup can be described as a one-parameter strongly continuous semigroup satisfying the intertwining relation (6.1). It is normalized iff it is contractive. This was the original definition of units going back to Powers and Arveson and it works for e_0-semigroups also. We call this description as the *operator picture* of units and the description as in Definition 6.1 as the *(dominated) map picture*.

Units of E_0-semigroups can also be thought of as some special parametrized families of vectors from the associated product system. As in Section 3, fix $a \in \mathcal{H}$ with $\|a\| = 1$, and consider the product system $\{\mathcal{P}_t : t \geq 0\}$ with a as ground vector, and unitary operators $U_{s,t} : \mathcal{P}_s \otimes \mathcal{P}_t \to \mathcal{P}_{s+t}$ as in Proposition 3.2. Then a normalized unit for θ is a family $v = \{v(t) : t \geq 0\}$ of vectors, $v(t) \in \mathcal{P}_t, v(0) = a, \|v(t)\| \leq 1$, such that $U_{s,t}(v(s) \otimes v(t)) = v(s + t)$, for $s, t \geq 0$ and the map $t \mapsto v(t)$, from $\mathbb{R}_+$ to $\mathcal{H}$ is continuous. This we call as the *vector picture* of units. For a unit $V = \{V_t : t \geq 0\}$ in the operator picture, the corresponding unit now is the family $v = \{v_t : t \geq 0\}$, defined by $v_t = V_t a, t \geq 0$. (As $\theta_t(|a\rangle\langle a|)V_t = V_t(|a\rangle\langle a|)a = V_t a = v_t$, $\mathcal{P}_t$ does contain v_t). The inverse map is also easy to describe [Bh2], in fact for v as above simply define V_t by

$$V_t z = \theta_t(|z\rangle\langle a|)v_t, \quad z \in \mathcal{H}, t \geq 0,$$

and verify all the required properties. So far we have three descriptions of units for E_0-semigroups. We will have another picture (which is valid even for quantum dynamical semigroups) later in this section. We can always opt for the most convenient description for the situation at hand.

Remark 6.2: If $\{G_t : t \geq 0\}$ is a contractive local cocycle of θ and $\{v_t : t \geq 0\}$ is a unit then $\{G_t v_t : t \geq 0\}$ is also a unit.

Proof: We make use of the operator picture and take $v_t = V_t a$ for a unit V and ground vector a. Now for $s, t \geq 0$, we have

$$G_s V_s G_t V_t = G_s \theta_s(G_t) V_s V_t = G_{s+t} V_{s+t}.$$

Also, for $Z \in \mathcal{B}(\mathcal{H})$,

$$\theta_t(Z) G_t V_t = G_t \theta_t(Z) V_t = G_t V_t Z.$$

Finally, strong operator continuity of $t \mapsto G_t V_t$ is clear as G_t, V_t are individually strongly continuous and they are norm-bounded. ∎

Now we study the behaviour of units under dilation. We assume the set up of Section 3, namely τ is a unital quantum dynamical semigroup of $\mathcal{B}(\mathcal{H}_0)$, and an E_0-semigroup θ of $\mathcal{B}(\mathcal{H})(\mathcal{H}_0 \subseteq \mathcal{H})$, is a dilation of τ, and $\hat{\tau}, \tilde{\tau}$ are associated minimal, induced semigroups. As before P denotes the projection of $\mathcal{H}$ on to $\mathcal{H}_0$.

Suppose $V = \{V_t; t \geq 0\}$ is a normalized unit of θ. As $\theta_t(P)$ is increasing, it is easily seen that $PV_tP = PV_t$ (that is, $V_t{}^*$ leaves $\mathcal{H}_0$ invariant). Then the compression $A_t = PV_tP = PV_t$, becomes a normalized unit of τ. Similar to Section 5, we would like to show that this compression map from units of θ to units of τ is always surjective, and is injective if θ is the minimal dilation of τ. It is not clear as to whether in general injectivity necessarily implies minimality in case of compression of units. It is the case for type I dilations as Corollary 7.7 shows.

Let $A = \{A_t : t \geq 0\}$ be a normalized unit of τ. We would like to construct normalized units of $\hat{\tau}, \tilde{\tau}, \theta$ such that their compression to $\mathcal{H}_0$ is A. We do this quite explicitly. For $(\underline{r}, \underline{Y}, u) \in \mathcal{N}$ (recall the definition of $\mathcal{N}$ from Section 3) with $r_1 \geq r_2 \geq \cdots r_k \geq t \geq r_{k+1} \geq \cdots r_n \geq 0$, set

$$\hat{A}_t^*\theta(\underline{r}, \underline{Y})u = \theta_{r_1-t}(Y_1) \cdots \theta_{r_k-t}(Y_k)A_{t-r_{k+1}}^* Y_{k+1} A_{r_{k+1}-r_{k+2}}^* \cdots Y_n A_{r_n}^* u \, (6.2)$$

Theorem 6.3: The family $\hat{A} = \{\hat{A}_t : t \geq 0\}$ defined by taking adjoints $\hat{A}_t^*$ as in (6.2) is the unique normalized unit of $\hat{\tau}$, such that its compression to $\mathcal{H}_0$ is A.

Proof : Consider the elementary semigroup $\alpha = \{\alpha_t : t \geq 0\}$, defined by $\alpha_t(X) = A_t X A_t^*, X \in \mathcal{B}(\mathcal{H}_0)$. We know that τ dominates α. Then making use of results of Section 5 we have a cocycle $\hat{G}$ of $\hat{\tau}$ such that the compression of $\hat{\tau}_t(\cdot)\hat{G}_t$ to $\mathcal{B}(\mathcal{H}_0)$ is α_t. Now for $(\underline{r}, \underline{Y}, u), (\underline{r}, \underline{Z}, v) \in \mathcal{N}$ from (5.1) we obtain

$$\langle \hat{A}_t^*\theta(\underline{r}, \underline{Y})u, \hat{A}_t^*\theta(\underline{r}, \underline{Z})v \rangle = \langle \theta(\underline{r}, \underline{Y})u, \hat{G}_t\theta(\underline{r}, \underline{Z})v \rangle.$$

It follows that $\hat{A}_t^*$ is a bounded operator, and $\hat{A}_t\hat{A}_t^* = \hat{G}_t$, for $t \geq 0$. From the very definition it is clear that $\{\hat{A}_t^* : t \geq 0\}$ is a one-parameter semigroup. Therefore $\hat{A} = \{\hat{A}_t : t \geq 0\}$ is also a semigroup. In particular, $\|\hat{A}_t\|^2 = \|\hat{G}_t\| \leq 1$.

The intertwining property for $\hat{A}$ can be verified by first observing $\hat{A}_t^* \theta_t(Z) = Z\hat{A}_t^*$, for rank one operators $Z = |\xi\rangle\langle\eta|$ with $\xi = \theta(\underline{p}, \underline{X})x$, $\eta = \theta(\underline{q}, \underline{W})w$, $(\underline{p}, \underline{X}, x), (\underline{q}, \underline{W}, w) \in \mathcal{N}$ and then extending the result to general $Z \in \mathcal{B}(\hat{\mathcal{H}})$. For $u, v \in \mathcal{H}_0, t \geq 0$

$$\langle u, \hat{A}_t v\rangle = \langle \hat{A}_t^* u, v\rangle = \langle A_t^* u, v\rangle = \langle u, A_t v\rangle.$$

Therefore the compression of $\hat{A}_t$ to $\mathcal{H}_0$ is A_t. Strong continuity of $\hat{A}$ follows from Appendix A, Proposition A.7. Finally, to show uniqueness observe that intertwining and semigroup property for units gives (for $(\underline{r}, \underline{Y}, u) \in \mathcal{N}$ with $r_k \geq t \geq r_{k+1}$)

$$\begin{aligned}
\hat{A}_t^* \hat{\tau}(\underline{r}, \underline{Y})u &= \hat{\tau}_{r_1-t}(Y_1) \cdots \hat{\tau}_{r_k-t}(Y_r) \hat{A}_t^* \hat{\tau}_{r_{k+1}}(Y_{k+1}) \cdots \hat{\tau}_{r_n}(Y_n)u \\
&= \hat{\tau}_{r_1-t}(Y_1) \cdots \hat{\tau}_{r_k-t}(Y_k) \hat{A}_{t-r_{k+1}}^* Y_{k+1} \hat{A}_{r_{k+1}}^* \hat{\tau}_{r_{k+2}} \cdots \hat{\tau}_{r_n}(Y_n)u \\
&= \hat{\tau}_{r_1-t}(Y_1) \cdots \hat{\tau}_{r_k-t}(Y_k) \hat{A}_{t-r_{k+1}}^* Y_{k+1} \hat{A}_{r_{k+1}-r_{k+2}}^* Y_{k+2} \cdots Y_n \hat{A}_{r_n}^* u.
\end{aligned}$$

As we want compression of $\hat{A}_{r_n}^*$ to $\mathcal{H}_0$ be $A_{r_n}^*$, we have $Y_n \hat{A}_{r_n}^* u = Y_n A_{r_n}^* u$. Repeating this argument we get

$$\hat{A}_t^* \hat{\tau}(\underline{r}, \underline{Y})u = \hat{\tau}_{r_1-t}(Y_1) \cdots \hat{\tau}_{r_k-t}(Y_k) A_{t-r_{k+1}}^* Y_{k+1} A_{r_{k+1}-r_{k+2}}^* \cdots Y_n A_{r_n}^* u.$$

But from (3.1) $\hat{\tau}(\underline{r}, \underline{Y})u = \theta(\underline{r}, \underline{Y})u$. So $\hat{A}_t^*$ must be given by (6.2). ∎

Fix a unit vector $a \in \mathcal{H}_0$. Considering the product system $\hat{\mathcal{P}}_t =$ range of $\hat{\tau}_t(|a\rangle\langle a|)$, the unit $\hat{A}$ has the vector picture $v = \{v_t : v_t \in \hat{\mathcal{P}}_t\}$ given by $v_t = \hat{A}_t a, t \geq 0$. The same v can be thought of as a unit for $\tilde{\tau}$ and θ as well. However we would like to continue to work with the operator picture of units. Recalling the definition of W_t from Section 3, note that

$$\hat{A}_t u = W_t(u \otimes v_t), \quad \text{for } u \in \hat{\mathcal{H}}.$$

Define $\tilde{A} = \{\tilde{A}_t : t \geq 0\}$ by

$$\tilde{A}_t z = W_t(z \otimes v_t), \quad \text{for } z \in \mathcal{H}.$$

Theorem 6.4: $\tilde{A}$ is the unique normalized unit of $\tilde{\tau}$ such that its compression to $\mathcal{H}_0$ is A. Moreover it is also a normalized unit of θ.

Proof : For $s, t \geq 0, \quad z \in \mathcal{H}$

$$\begin{aligned}
\tilde{A}_s \tilde{A}_t z &= W_s(W_t(z \otimes v_t) \otimes v_s) = W_s(W_t \otimes 1_{\mathcal{P}_s})(z \otimes v_t \otimes v_s) \\
&= W_{s+t}(1 \otimes U_{t,s})(z \otimes v_t \otimes v_s) = W_{s+t}(z \otimes v_{s+t}) \\
&= \tilde{A}_{s+t} z,
\end{aligned}$$

as $\{v_t\}$ is a unit for the product system $\{\hat{\mathcal{P}}_t, \hat{U}_{s,t}\}$. Further $\|v_t\| \leq 1, \quad \forall t$ implies $\|\tilde{A}_t\| \leq 1$. Now for $Z \in \mathcal{B}(\mathcal{H}), z \in \mathcal{H}$,

$$\tilde{\tau}_t(Z)\tilde{A}_t z = W_t(Z \otimes 1_{\hat{\mathcal{P}}_t})W_t^* W_t(z \otimes v_t) = W_t(Zz \otimes v_t) = \tilde{A}_t Zz$$

Therefore $\tilde{A}$ is a unit of $\tilde{\tau}$. As $\tilde{A}_t z = \hat{A}_t z$ for $z \in \hat{\mathcal{H}}$, the compression of $\tilde{A}$ to $\hat{\mathcal{H}}, \mathcal{H}_0$ are $\hat{A}, A$ respectively.

Now if $\tilde{B} = \{\tilde{B}_t : t \geq 0\}$ is a unit of $\tilde{\tau}$. We have

$$\tilde{\tau}_t(Z)\tilde{B}_t z = W_t(Z \otimes 1_{\hat{\mathcal{P}}_t})W_t^* \tilde{B}_t z = \tilde{B}_t Zz, \quad \text{for } z \in \mathcal{H},$$

proving $\tilde{B}_t z = z \otimes w_t$, for some $w_t \in \hat{\mathcal{P}}_t$ (An elementary argument shows that w_t can not depend upon z). If the compression of $\tilde{B}$ to $\mathcal{H}_0$ is A, its

COCYCLES OF CCR FLOWS

compression to $\hat{\mathcal{H}}$ must be $\hat{A}$ by uniqueness in Theorem 6.3. This gives $w_t = v_t$, in other words $\tilde{B} = \tilde{A}$. The final statement in the Theorem follows trivially as any unit of $\tilde{\tau}$ is a unit of θ by considering dominated map picture.

∎

In view of Theorem 6.4 if τ has units then any dilation θ of it also has units. Conversely, if θ has units then τ has by compression. Moreover the compression map maps normalized units $\mathcal{U}_0(\theta)$ surjectively to normalized units $\mathcal{U}_0(\tau)$. From uniqueness mentioned in Theorem 6.3, we also see that the compression map is injective on normalized units for minimal dilations. Suppose $\{V_t\}$ is a unit of θ and $\{E_t\}$ is the projection cocycle $E_t = \tilde{\tau}_t(1), \forall t$, then $\{E_t V_t\}$ is the unique unit of $\tilde{\tau}$ having the property that it has same compression as $\{V_t\}$ to $\hat{\mathcal{H}}$. This remark is to be compared with Theorem 5.4.

Let $\mathcal{H}_0^+$ be the Hilbert space $\mathbb{C} \oplus \mathcal{H}_0$. For $c \in \mathbb{C}, x, y \in \mathcal{H}_0$ and $X \in \mathcal{B}(\mathcal{H}_0)$, we have a bounded operator

$$X^+ = \begin{bmatrix} c & y^* \\ x & X \end{bmatrix} \quad \text{on } \mathcal{H}_0^+$$

defined by $X^+(d \oplus u) = (cd + \langle y, u \rangle) \oplus (dx + Xu)$, for $d \oplus u \in \mathcal{H}_0^+$.

Definition 6.5: Let τ be a unital quantum dynamical semigroup of $\mathcal{B}(\mathcal{H})$. A quantum dynamical semigroup τ^+ of $\mathcal{B}(\mathbb{C} \oplus \mathcal{H}_0)$ is said to be a 1 - dimensional unital extension of τ if

$$\tau_t^+ \left(\begin{bmatrix} c & 0 \\ 0 & X \end{bmatrix} \right) = \begin{bmatrix} c & 0 \\ 0 & \tau_t(X) \end{bmatrix}$$

for every $c \in \mathbb{C}, X \in \mathcal{B}(\mathcal{H}_0)$.

Theorem 6.6: There is a 1-1 correspondence between normalized units of τ and 1-dimensional unital extensions of τ.

Proof : Let $A = \{A_t : t \geq 0\}$ be a normalized unit of τ. Take $\mathcal{H}_0^+ = \mathbb{C} \oplus \mathcal{H}_0$ and define a quantum dynamical semigroup τ^+ of $\mathcal{B}(\mathcal{H}_0^+)$ by

$$\tau_t^+\left(\begin{bmatrix} c & y^* \\ x & X \end{bmatrix}\right) = \begin{bmatrix} c & (A_t y)^* \\ A_t x & \tau_t(X) \end{bmatrix}$$

for $c \in \mathbb{C}, x, y \in \mathcal{H}_0, X \in \mathcal{B}(\mathcal{H}_0)$. As we have

$$\tau_t^+\left(\begin{bmatrix} c & y^* \\ x & X \end{bmatrix}\right) = \begin{bmatrix} 1 & 0 \\ 0 & A_t \end{bmatrix}\begin{bmatrix} c & y^* \\ x & X \end{bmatrix}\begin{bmatrix} 1 & 0 \\ 0 & A_t^* \end{bmatrix} + \begin{bmatrix} 0 & 0 \\ 0 & \tau_t(X) - A_t X A_t^* \end{bmatrix},$$

and τ_t dominates $A_t(\cdot)A_t^*$, τ_t^+ is completely positive. Semigroup property of τ^+ is obvious. Continuity is also equally obvious.

In the converse direction, suppose $\bar{\tau}$ is a 1-dimensional unital extension of τ. Define $A_t : \mathcal{H}_0 \to \mathcal{H}_0$ by

$$A_t u = \bar{\tau}_t(|u\rangle\langle e|)e, \quad u \in \mathcal{H}_0$$

where e is the vector $(1 \oplus 0)$ in $\mathbb{C} \oplus \mathcal{H}_0$. To see that A_t actually maps $\mathcal{H}_0$ to $\mathcal{H}_0$, recall that if α is a contractive completely positive map $\alpha(Z^*)\alpha(Z) \leq \alpha(Z^*Z)$ for any Z. Thus for $u \in \mathcal{H}_0$

$$
\begin{aligned}
|\langle e, \bar{\tau}_t(|u\rangle\langle e|)e\rangle|^2 &= |\langle \bar{\tau}_t(|e\rangle\langle u|)e, e\rangle|^2 \\
&\leq \|\bar{\tau}_t(|e\rangle\langle u|)e\|^2 \\
&\leq \langle e, \bar{\tau}_t(|u\rangle\langle u|)e\rangle
\end{aligned}
$$

$$= \langle e, \tau_t(|u\rangle\langle u|)e \rangle$$

$$= 0.$$

Now we claim that $\bar{\tau}_t(|u\rangle\langle e|) = |A_t u\rangle\langle e|$, for $u \in \mathcal{H}_0$. This is clear as $\bar{\tau}_t(|u\rangle\langle e|)e = A_t u$ by definition of A_t, and for any $v \in \mathcal{H}_0$

$$
\begin{aligned}
\|\bar{\tau}_t(|u\rangle\langle e|)v\|^2 &= \langle v, \bar{\tau}_t(|e\rangle\langle u|)\bar{\tau}_t(|u\rangle\langle e|)v \rangle \\
&\leq \|u\|^2 \langle v, \bar{\tau}_t(|e\rangle\langle e|)v \rangle \\
&= \|u\|^2 \langle v, (|e\rangle\langle e|)v \rangle \\
&= 0.
\end{aligned}
$$

It follows that

$$
\bar{\tau}_t \left(\begin{bmatrix} c & y^* \\ x & X \end{bmatrix} \right) = \begin{bmatrix} c & (A_t y)^* \\ A_t x & \tau_t(X) \end{bmatrix}
$$

and $A = \{A_t : t \geq 0\}$ is a semigroup of contractions. Finally to see that τ_t dominates $A_t(\cdot)A_t^*$, consider the minimal dilation $\bar{\theta}$ of $\bar{\tau}$. From Theorem 6.4 of [Bh2], we know that the vector $e = 1 \oplus 0$, generates a unit V of $\bar{\theta}$, given by

$$V_t z = \bar{\theta}_t(|z\rangle\langle e|)e$$

for all z in the minimal dilation space $\bar{\mathcal{H}}$ of $\bar{\tau}$. Compressing V to $\mathbb{C} \oplus \mathcal{H}_0$ we get a semigroup $B = \{B_t : t \geq 0\}$ such that $\bar{\tau}_t(\cdot)$ dominates $B_t(\cdot)B_t^*$. Now for $x = c \oplus u, y = d \oplus v$, in $\mathbb{C} \oplus \mathcal{H}_0$, we have

$$
\begin{aligned}
\langle x, V_t y \rangle &= \langle x, \bar{\theta}_t(|y\rangle\langle e|)e \rangle \\
&= \langle x, \bar{\tau}_t(|y\rangle\langle e|)e \rangle \\
&= \langle x, d \oplus A_t v \rangle \\
&= \bar{c}d + \langle u, A_t v \rangle.
\end{aligned}
$$

That is,

$$B_t = \begin{bmatrix} 1 & 0 \\ 0 & A_t \end{bmatrix}.$$

Then for $Z = \begin{bmatrix} c & y^* \\ x & X \end{bmatrix}$ in $\mathcal{B}(\mathbb{C} \oplus \mathcal{H}_0)$

$$\begin{aligned}
\bar{\tau}_t(Z) - B_t Z B_t^* &= \begin{bmatrix} c & (A_t y)^* \\ A_t x & \tau_t(X) \end{bmatrix} - \begin{bmatrix} 1 & 0 \\ 0 & A_t \end{bmatrix} \begin{bmatrix} c & y^* \\ x & X \end{bmatrix} \begin{bmatrix} 1 & 0 \\ 0 & A_t^* \end{bmatrix} \\
&= \begin{bmatrix} 0 & 0 \\ 0 & \tau_t(X) - A_t X A_t^* \end{bmatrix}.
\end{aligned}$$

Hence τ_t dominates $A_t(\cdot)A_t^*$. It is clear that the mapping we constructed between units and extensions of τ are inverses of each others. ∎

In view of Theorem 6.6 we can describe normalized units as 1-dimensional unital extensions. This we call as *extension picture* of units. Davies [Da1-4] did pioneering work in studying quantum dynamical semigroups with unbounded generators. For his analysis it was essential that these semigroups have 1-dimensional unital extensions. He exploits the fact that these extensions have pure, normal invariant states. Davies comments that (page 431 [Da4]) we don't know whether such extensions exist for all semigroups. Now that we have connected this to the question of existence of units the problem is easily resolved. Powers [Po2] has already exhibited one E_0-semigroup having no unit. So quantum dynamical semigroups with no 1-dimensional unital extensions do exist. As in E_0-semigroup theory semigroups which have units seem to behave better. It is to be noted that so far most papers studying unbounded generators of quantum dynamical semigroups (See for example [BS], [Ch], [CF], [Fa1-2], [Mo]) have used the set up of Davies. So almost all quantum dynamical semigroups appearing there have atleast one unit. In fact usually these semigroups are obtained by adding a series of completely positive maps to some elementary semigroup. For the same reason Feller perturbed quantum dynamical semigroups of [BP1] also have units. However it is perhaps possible that minimal dilations of some of these semigroups are of type II and not of type I. That will be interesting.

7 Cocycle computation for CCR flows

Let $\mathcal{K}$ be a complex separable Hilbert space. Let $\mathcal{H} = \Gamma(L^2(\mathbb{R}_+, \mathcal{K}))$ be the Boson Fock space over the Hilbert space $L^2(\mathbb{R}_+, \mathcal{K})$, of $\mathcal{K}$ valued square integrable functions on $\mathbb{R}_+$. For $u \in L^2(\mathbb{R}_+, \mathcal{K})$, we have the exponential vector $e(u) \in \Gamma(L^2(\mathbb{R}_+, \mathcal{K}))$ defined by

$$e(u) = 1 \oplus u \oplus \frac{u^{\otimes 2}}{\sqrt{2!}} \oplus \cdots,$$

satisfying $\langle e(u), e(v) \rangle = e^{\langle u, v \rangle}$, for $u, v \in L^2(\mathbb{R}_+, \mathcal{K})$.

For $t \geq 0$, let $\chi_{t]}, \chi_{[t}$ be indicator functions of $[0, t], [t, \infty)$ respectively. The Hilbert space $\mathcal{H}$ decomposes naturally as $\mathcal{H}_{t]} \otimes \mathcal{H}_{[t}$, where $\mathcal{H}_{t]} = \Gamma(L^2([0, t], \mathcal{K}))$, $\mathcal{H}_{[t} = \Gamma(L^2([t, \infty), \mathcal{K}))$, by identifying $e(u) \in \mathcal{H}$ with $e(u\chi_{t]}) \otimes e(u\chi_{[t})$. Let S_t be the shift operator on $L^2(\mathbb{R}_+, \mathcal{K})$ defined by

$$S_t v(s) = \begin{cases} v(s - t) & s \geq t \\ 0 & 0 \leq s < t \end{cases}$$

Define unitary operators $W_t : \mathcal{H} \otimes \mathcal{H}_{t]} \to \mathcal{H}$ by setting

$$W_t(e(v) \otimes e(u)) = e(u + S_t v),$$

for $u \in L^2([0, t], \mathcal{K}) \subseteq L^2(\mathbb{R}_+, \mathcal{K}), v \in L^2(\mathbb{R}_+, \mathcal{K})$. Now for $Z \in \mathcal{B}(\mathcal{H})$, define $\gamma_t(Z) \in \mathcal{B}(\mathcal{H})$ by

$$\gamma_t(Z) = W_t(Z \otimes I_{t]})W_t^*$$

($I_{t]}$ being the identity operator on $\mathcal{H}_{t]}$). It is easily verified that $\gamma = \{\gamma_t : t \geq 0\}$ is an E_0-semigroup of $\mathcal{B}(\mathcal{H})$ satisfying

$$\gamma_t(|e(u)\rangle\langle e(v)|) = I_{t]} \otimes |e(S_t u)\rangle\langle e(S_t v)| \quad \text{for} \quad u, v \in L^2(\mathbb{R}_+, \mathcal{K}).$$

Taking the vacuum $e(0)$ as ground vector the product system $\mathcal{P}_t$ becomes $\Gamma(L^2([0,t],\mathcal{K})$. Arveson index of $\gamma = \gamma^{\mathcal{K}}$ is same as $\dim(\mathcal{K})$. We shall call γ as CCR flow of index $\dim(\mathcal{K})$. We want to determine all quantum dynamical semigroups dominated by γ. By Theorem 4.3, this amounts to computing all positive, contractive, local cocycles of γ. We will show that this set is parameterized by triples (A, x, q) where A is a bounded operator in $\mathcal{B}(\mathcal{K})$, x is a vector in $\mathcal{K}$ and q is a scalar in $\mathbb{R}$, such that

(a) $0 \le A \le I$;

(b) $x \in \operatorname{range} (I - A)^{1/2}$;

(c) $q \ge q_0(A, x)$, where $q_0(A, x) = \inf\{\|a\|^2 : x = (I - A)^{1/2}a\}$.

Let $\mathcal{D}_{\mathcal{K}}$ be the set of all such triples. For any $(A, x, q) \in \mathcal{B}(\mathcal{K}) \times \mathcal{K} \times \mathbb{C}$, we can define a bounded operator $[A, x, q]$ on $\mathcal{K}^+ = \mathbb{C} \oplus \mathcal{K}$, by taking

$$[A, x, q] = \begin{bmatrix} -q & x^* \\ x & A \end{bmatrix}$$

that is, $[A, x, q](c \oplus y) = (-qc + \langle x, y \rangle) \oplus (cx + Ay)$, for $(c \oplus y)$ in $\mathcal{K}^+$.

Lemma 7.1 : (i) Suppose a pair $(A, x) \in \mathcal{B}(\mathcal{K}) \times \mathcal{K}$ satisfies (a), (b). Then there exists unique $a_0 \in \mathcal{K}$ with $x = (I - A)^{1/2}a_0$ such that $q_0(A, x) = \|a_0\|^2$. If $x = (I - A)b$ for some $b \in \mathcal{K}$ then $a_0 = (I - A)^{1/2}b$.

(ii) A triple $(A, x, q) \in \mathcal{B}(\mathcal{K}) \times \mathcal{K} \times \mathbb{C}$ is in $\mathcal{D}_{\mathcal{K}}$ if and only if $A \ge 0$ and $[A, x, q] \le 1_{\mathcal{K}}$, where $1_{\mathcal{K}}$ is the orthogonal projection of $\mathcal{K}^+$ onto $\mathcal{K}$.

Proof : We have $\overline{\text{range}}\,(I-A) = \overline{\text{range}}\,(I-A)^{1/2}$, and $(\overline{\text{range}}\,(I-A))^{\perp} =$ $\ker\,(I-A) = \ker\,(I-A)^{1/2}$. Let Q be the projection onto $\overline{\text{range}}\,(I-A)$. Now if $x = (I-A)^{1/2}c$ for some c, taking $a_0 = Qc$, we observe that $\{a : x = (I-A)^{1/2}a\} = \{a_0 + z : z \in \ker\,(I-A)^{1/2}\}$. From this (i) follows quite easily.

For proving (ii) recall that ([FF] page 547) a block operator matrix

$$\begin{bmatrix} P & Q^* \\ Q & R \end{bmatrix}$$

is positive if and only if $P, R \geq 0$ and there exists a contraction $S :$ $\overline{\text{range}}\,P \to \overline{\text{range}}\,R$, such that $Q = R^{1/2}SP^{1/2}$. In our case,

$$1_{\mathcal{K}} - [A, x, q] = \begin{bmatrix} q & -x^* \\ -x & (I-A) \end{bmatrix}$$

is positive if and only if $q \geq 0, (I-A) \geq 0$ and $x = (I-A)^{1/2}sq^{1/2}$, for some $s \in \overline{\text{range}}\,(I-A), \|s\| \leq 1$. We have $sq^{\frac{1}{2}} = a_0$ and now $\|s\|^2 \leq 1$ iff $q \geq \|a_0\|^2 = q_0(A, x)$.   ∎

We define a partial order '$\leq$' on $\mathcal{D}_{\mathcal{K}}$ by setting $(A_1, x_1, q_1) \leq (A_2, x_2, q_2)$ iff $[A_1, x_1, q_1] \leq [A_2, x_2, q_2]$. Here after we need not really distinguish between triple (A, x, q) and operator $[A, x, q]$.

For $(A, x, q) \in \mathcal{D}_{\mathcal{K}}$, we define a cocycle $G = G(A, x, q)$ of γ by setting

$$G_t e(v) = e^{-qt + \langle x\chi_{t]}, v\chi_{t]}\rangle} e(x\chi_{t]} + Av\chi_{t]} + v\chi_{[t}), \qquad (7.1)$$

for $t \geq 0, v \in L^2(\mathbb{R}_+, \mathcal{K})$. Here $x\chi_{t]}, Av\chi_{t]}$ denote the vectors

$$x\chi_{t]}(s) = \begin{cases} x & 0 \leq s < t \\ 0 & s \geq t \end{cases} \qquad Av\chi_{t]}(s) = \begin{cases} Av(s) & 0 \leq s < t \\ 0 & s \geq t \end{cases}$$

in $L^2([0,t],\mathcal{K}) \subseteq L^2(\mathbb{R}_+,\mathcal{K})$. Similarly the vector $v\chi_{[t}$ is the truncation of v to $L^2([t,\infty),\mathcal{K})$. Of course, we need to show that G_t extends to a contraction operator on $\mathcal{H}$. Norms of such operators have been computed by Araki and Woods in [AW]. But we will not be using their results.

If x is in range $(I-A)$, say $x = (I-A)b$, for some $b \in \mathcal{K}$ then G_t can be written as

$$G_t = e^{-(q-q_0)t} W(b\chi_{t]})\Gamma_t(A)W(-b\chi_{t]}) \tag{7.2}$$

with $q_0 = \langle b,(I-A)b \rangle$, where $W(u)$ is the Weyl operator ([Pa], page 135) defined by

$$W(u)e(v) = e^{-\frac{1}{2}\|u\|^2 - \langle u,v \rangle}e(u+v) \qquad \text{for } u,v \in L^2(\mathbb{R}_+,\mathcal{K}),$$

and $\Gamma_t(A)$ is the second quantization operator ([Pa], pages 136, 150) defined by

$$\Gamma_t(A)e(v) = e(Av\chi_{t]} + v\chi_{[t}) \qquad \text{for } v \in L^2(\mathbb{R}_+,\mathcal{K}).$$

(Strictly speaking we are having a second quantized operator on $\mathcal{B}(\mathcal{H}_{t]})$ ampliated to $\mathcal{B}(\mathcal{H})$). As Weyl operators are unitaries with $W(u)^* = W(-u)$, and $0 \le \Gamma_t(A) \le I$, with $\|\Gamma_t(A)\| \equiv 1$, for $0 \le A \le I$, we deduce that for $x \in$ range $(I-A)$, G_t is a well-defined operator, $0 \le G_t \le I$, with $\|G_t\| = e^{-(q-q_0)t}$. ($q_0 = q_0(A,x)$). Note that if $\dim(\mathcal{K}) < \infty$ then range $(I-A)^{1/2} =$ range $(I-A)$, so every $G_t(A,x,q)$ is given by (7.2). If $\dim \mathcal{K} = \infty$, and x is in range $(I-A)^{1/2}$ but is not in range $(I-A)$, then perhaps we can approximate x by suitable $x_n \in$ range $(I-A)$, so that G_t is a strong limit of positive contractions. But instead we prove contractivity of G_t's very directly. For this we need the notion of conditional positive definiteness.

Definition 7.2: An $n \times n$ matrix $B = [b_{ij}]$ is said to be conditionally positive definite if it is self-adjoint and $\sum_{i,j=1}^{n} \overline{c_i} c_j b_{ij} \geq 0$ for all $c_1, c_2, \cdots, c_n \in \mathbb{C}$ with $\sum c_i = 0$.

The following lemma seems to be well-known.

Lemma 7.3: For $n \times n$ complex matrix B, the following are equivalent

(i) $[e^{tb_{ij}}] \geq 0$ for all $t \geq 0$;

(ii) B is conditionally positive definite;

(iii) $b_{ij} = b + \overline{b_i} + b_j + c_{ij}$, for some $b \in \mathbb{R}, b_1, b_2, \ldots b_n \in \mathbb{C}$ and $C = [c_{ij}] \geq 0$.

Suppose B is conditionally positive definite then for any matrix D, $[e^{tb_{ij}}] \leq [e^{td_{ij}}]$ for all t if and only if $B \leq D$.

Proof : See for example [Gu], [PSc] for equivalence of (i) to (iii). The last statement is elementary. ∎

Lemma 7.4: For $(A, x, q), (A', x', q') \in \mathcal{D}_{\mathcal{K}}, G_t(A, x, q), G_t(A', x', q')$ are positive contractions. Moreover $G_t(A, x, q) \leq G_t(A', x', q')$ iff $(A, x, q) \leq (A', x', q')$.

Proof : First note that if $(A', x', q') = (I, 0, 0)$, then $G_t(A', x', q')$ is nothing but the identity operator on $\mathcal{H}$. So the contractivity of G_t would follow from the second statement in the Lemma. As for positivity of $G_t = G_t(A, x, q)$, it is clear from $(iii) \Rightarrow (i)$ of Lemma 7.3 as,

$$\langle \sum_i c_i e(v_i), G_t \sum_j c_j e(v_j) \rangle$$
$$= \sum_{i,j} \overline{c_i} c_j e^{-qt + \langle x\chi_{t]}, v_j \rangle + \langle v_i, x\chi_{t]} \rangle + \langle v_i\chi_{t]}, Av_j\chi_{t]} \rangle + \langle v_i\chi_{[t}, v_j\chi_{[t} \rangle}$$

for $c_1, \ldots, c_n \in \mathbb{C}, v_i, \ldots, v_n \in L^2(\mathbb{R}_+, \mathcal{K})$. This conclusion needs nothing more than positivity of A. Observe that from Lemma 7.3,

$$\langle \sum_i c_i e(v_i), G_t \sum_j c_j e(v_j) \rangle \leq \langle \sum_i c_i e(v_i), G'_t \sum_j c_j e(v_j) \rangle$$

iff $[a_{ij}(t)] \geq 0$ where

$$a_{ij}(t) = -(q' - q)t + \langle (x' - x)\chi_{t]}, v_j \rangle + \langle v_i, (x' - x)\chi_{t]} \rangle$$
$$+ \langle v_i\chi_{t]}, (A' - A)v_j\chi_{t]} \rangle.$$

Now suppose $(A, x, q) \leq (A', x', q')$. We have

$$\begin{bmatrix} -q & x^* \\ x & A \end{bmatrix} \leq \begin{bmatrix} -q' & (x')^* \\ x' & A' \end{bmatrix}$$

or

$$\begin{bmatrix} -(q' - q) & (x' - x)^* \\ x' - x & A' - A \end{bmatrix} \geq 0$$

From Lemma 7.1 (ii) (or its proof), this means that $A' - A \geq 0$, there exists $a \in \mathcal{K}$ such that $(A' - A)^{1/2}a = x' - x$ and $\|a\|^2 \leq -(q' - q)$. Hence taking

$$b_{ij}(s) = -(q' - q) + \langle (A' - A)^{1/2}a, v_j(s) \rangle + \langle v_i(s), (A' - A)^{1/2}a \rangle$$
$$+ \langle (A' - A)^{1/2}v_i(s), (A' - A)^{1/2}v_j(s) \rangle$$
$$= [-(q' - q) - \|a\|^2] + \langle a + (A' - A)^{1/2}v_i(s), a + (A' - A)^{1/2}v_j(s) \rangle,$$

clearly,

$$[a_{ij}(t)] = [\int_0^t b_{ij}(s)ds]$$

forms a positive definite matrix. In the converse direction, if $G_t(A, x, q) \leq G_t(A', x', q')$, we have $[a_{ij}(t)] \geq 0$, in particular $a_{11}(t) \geq 0$. Taking $v_1 = y\chi_{t]}$, for arbitrary vectors y in $\mathcal{K}$, we have

$$-(q' - q)t + \langle (x' - x), y \rangle t + \langle y, (x' - x) \rangle t + \langle y, (A' - A)y \rangle t \geq 0$$

For $c \in \mathbb{C}$ multiplying by $|c|^2/t$, and taking $z = cy$ we get

$$-|c|^2(q' - q) + \bar{c}\langle (x' - x), z \rangle + c\langle z, (x' - x) \rangle + \langle z, (A' - A)z \rangle \geq 0,$$

that is,

$$\left\langle \begin{pmatrix} c \\ z \end{pmatrix}, \begin{bmatrix} -(q' - q) & (x' - x)^* \\ (x' - x) & (A' - A) \end{bmatrix} \begin{pmatrix} c \\ z \end{pmatrix} \right\rangle \geq 0.$$

The case, $c = 0, z \neq 0$ is not covered by this argument, but that can easily be taken care of by taking limits. Therefore we conclude that $(A, x, q) \leq (A', x', q')$.   ∎

Theorem 7.5: A family $G = \{G_t : t \geq 0\}$ of bounded operators on $\mathcal{B}(\Gamma(L^2(\mathbb{R}_+, \mathcal{K})))$ is a positive, contractive, local cocycle for CCR flow $\gamma = \gamma^{\mathcal{K}}$, if and only if $G = G(A, x, q)$ as given by (7.1) for some $(A, x, q) \in \mathcal{D}_{\mathcal{K}}$. Furthermore this correspondence between $\mathcal{D}_\gamma$ and $\mathcal{D}_{\mathcal{K}}$ respects the partial orders.

Proof: Given $(A, x, q) \in \mathcal{D}_{\mathcal{K}}$, we have already shown that $G_t = G_t(A, x, q)$ defined by (7.1) are positive contractions. Continuity of $t \mapsto G_t$ in strong operator topology, and local cocycle property are easy to verify. We have also seen that $(A, x, q) \mapsto G(A, x, q)$ respects the partial orders.

Now consider any positive, contractive, local cocycle G of γ. We want to obtain $(A, x, q) \in \mathcal{D}_{\mathcal{K}}$ such that $G = G(A, x, q)$. By Arveson ([Ar1], Theorem 4.7), any unit $u = \{u(t)\}$ of γ has the form

$$u(t) = e^{-pt} e(y\chi_{t]}) \text{ for some } (p, y) \in \mathbb{C} \times \mathcal{K}.$$

(We are considering the vector picture of units taking the vacuum vector $e(0)$ as ground vector.) We know that G_t sends units to units (Remark 6.2). Hence, for $y \in \mathcal{K}$,

$$G_t e(y\chi_t) = e^{-p_y t} e(B_y \chi_{t]})$$

for some $(p_y, B_y) \in \mathbb{C} \times \mathcal{K}$. We claim that $y \mapsto B_y - B_0$ is linear. We have

$$\langle e(0), G_t e(y\chi_{t]}) \rangle = e^{-p_y t} = \langle G_t e(0), e(y\chi_{t]}) \rangle = e^{-\overline{p_0} t + \langle B_0, y \rangle t}, \quad \text{for } t \geq 0.$$

Therefore, $-p_y = -\overline{p}_0 + \langle B_0, y \rangle$. In particular, p_0 is real. So we have

$$\langle B_0, y \rangle = p_0 - p_y. \tag{7.3}$$

Now for $y, z \in \mathcal{K}$ $\langle e(z\chi_{t]}), G_t e(y\chi_{t]}) \rangle = e^{-p_y t + \langle z, B_y \rangle t}$ and $\langle G_t e(z\chi_{t]}), e(y\chi_{t]}) \rangle = e^{-\overline{p}_z t + \langle B_z, y \rangle t}$, for all t. Hence

$$
\begin{aligned}
-p_y + \langle z, B_y \rangle &= -\overline{p}_z + \langle B_z, y \rangle, \\
\langle B_0, y \rangle - p_0 + \langle z, B_y \rangle &= \langle z, B_0 \rangle - p_0 + \langle B_z, y \rangle, \\
\langle z, B_y - B_0 \rangle &= \langle B_z - B_0, y \rangle.
\end{aligned}
$$

Thus for $c \in \mathbb{C}, w, y, z \in \mathcal{K}$,

$$
\begin{aligned}
\langle z, B_{cy+w} - B_0 \rangle &= \langle B_z - B_0, cy + w \rangle \\
&= c \langle B_z - B_0, y \rangle + \langle B_z - B_0, w \rangle \\
&= c \langle z, B_y - B_0 \rangle + \langle z, B_w - B_0 \rangle.
\end{aligned}
$$

This shows that the map $y \mapsto B_y - B_0$ is a linear map on $\mathcal{K}$. Set $Ay = B_y - B_0$, for $y \in \mathcal{K}$. Take $x = B_0, q = p_0$. Then we have $-p_y = -q + \langle x, y \rangle$ and $B_y = Ay + x$. Hence

$$G_t e(y\chi_{t]}) = e^{-qt + \langle x\chi_{t]}, y\chi_{t]} \rangle} e(x\chi_{t]} + Ay\chi_{t]}), \qquad (7.4)$$

for $y \in \mathcal{K}, t \geq 0$. Fix $y \in \mathcal{K}$ and consider the function

$$f(t) = \langle e(y\chi_{t]}) - e(0), G_t(e(y\chi_{t]}) - e(0)) \rangle \geq 0.$$

We have $f(0) = 0$, and

$$f(t) = e^{-qt + \langle x, y \rangle t + \langle y, x \rangle t + \langle y, Ay \rangle t} - e^{-qt + \langle x, y \rangle t} - e^{-qt + \langle y, x \rangle t} + e^{-qt}.$$

Differentiating at $t = 0$, we get $f'(0) = -q + \langle x, y \rangle + \langle y, x \rangle + \langle y, Ay \rangle - (-q + \langle x, y \rangle) - (-q + \langle y, x \rangle) - q = \langle y, Ay \rangle \geq 0$. Hence $A \geq 0$. Further as G_t is a contraction, for $y \in \mathcal{K}, t \geq 0$

$$\langle e(y\chi_{t]}), G_t e(y\chi_{t]}) \rangle \leq \langle e(y\chi_{t]}), e(y\chi_{t]}) \rangle.$$

This yields, $e^{-qt + \langle x, y \rangle t + \langle y, x \rangle t + \langle y, Ay \rangle t} \leq e^{\langle y, y \rangle t}$ or

$$q - \langle x, y \rangle - \langle y, x \rangle + \langle y, (I - A)y \rangle \geq 0.$$

Now as in the proof of Lemma 7.4 (multiplying by $|c|^2$, taking $z = cy$, etc.) we obtain

$$\begin{bmatrix} q & -x^* \\ -x & (I - A) \end{bmatrix} \geq 0.$$

Therefore $(A, x, q) \in \mathcal{D}_{\mathcal{K}}$. As G_t's are local and form a cocycle, from (7.4) we get

$$G_t e\left(\sum_{i=1}^{k} y_i \chi_{[s_{i-1}, s_i]} + v\chi_{[t,\infty)} \right)$$

$$= e^{-qt + \langle x\chi_t, \sum_{i=1}^{k} y_i \chi_{[s_{i-1}, s_i]} + v\chi_{[t,\infty)} \rangle} \cdot e\left(x\chi_{t]} + A\left(\sum_{i=1}^{k} y_i \chi_{[s_{i-1}, s_i]} \right) + v\chi_{[t,\infty)} \right)$$

for $0 = s_0 < s_1 < \cdots < s_{k-1} < s_k = t, y_1, \ldots, y_k \in \mathcal{K}, v \in L^2(\mathbb{R}_+, \mathcal{K})$. Now as step functions are dense in $L^2(\mathbb{R}_+, \mathcal{K})$ we obtain the required form for G_t. ∎

Let $\mathcal{E}_{\mathcal{K}}$ denote the set of projection cocycles in $\mathcal{D}_{\mathcal{K}}$.

Theorem 7.6: The partially ordered set $\mathcal{E}_{\mathcal{K}}$ is given by

$$\mathcal{E}_{\mathcal{K}} = \{G(Q, x, \|x\|^2) : Q \text{ is a projection in } \mathcal{B}(\mathcal{K}), x \in \mathcal{K}, \text{ with } Qx = 0\}.$$

with partial order $(Q, x, \|x\|^2) \leq (R, y, \|y\|^2)$ iff $Q \leq R$ and $(I - R)x = y$.

Proof: Suppose $G_t = G_t(A, x, q)$ is a projection. Then we have

$$\begin{aligned}
G_t^2 e(v) &= e^{-2qt + \langle x\chi_{t]}, v\rangle + \langle x\chi_{t]}, x\chi_{t]} + Av\chi_{t]}\rangle} e(x\chi_{t]} + Ax\chi_{t]} + A^2 v\chi_{t]} + v\chi_{[t}) \\
&= G_t e(v) \\
&= e^{-qt + \langle x\chi_{t]}, v\rangle} e(x\chi_{t]} + Av\chi_{t]} + v\chi_{[t}).
\end{aligned}$$

Hence $x\chi_{t]} + Ax\chi_{t]} + A^2 v\chi_{t]} = x\chi_{t]} + Av\chi_{t]}, \quad \forall v \in L^2(\mathbb{R}_+, \mathcal{K})$. In particular, $\forall y \in \mathcal{K}, Ax = Ay - A^2 y$. Only way this is possible is that $Ax = 0$, and $Ay = A^2 y, \quad \forall y \in \mathcal{K}$. That is, A is a projection, and $x \in$ range $(I - A)$. Comparing the scalars in the equation above,

$$-2qt + \langle x\chi_{t]}, v\rangle + \langle x\chi_{t]}, x\chi_{t]} + Av\chi_{t]}\rangle = -qt + \langle x\chi_{t]}, v\rangle$$

Hence, $-qt + \|x\|^2 t = 0$, or $q = \|x\|^2 = q_0(A, x)$. This proves the first part. Now for $G(Q, x, \|x\|^2), G(R, y, \|y\|^2)$ in $\mathcal{E}_{\mathcal{K}}$. We have $G(Q, x, \|x\|^2) \leq G(R, y, \|y\|^2)$ iff

$$\begin{bmatrix} -\|x\|^2 & x^* \\ x & Q \end{bmatrix} \leq \begin{bmatrix} -\|y\|^2 & y^* \\ y & R \end{bmatrix},$$

iff

$$\left[\begin{array}{cc} \|x\|^2 - \|y\|^2 & (y-x)^* \\ y-x & R-Q \end{array} \right] \geq 0,$$

The last inequality holds iff $R-Q \geq 0$, and $(R-Q)(y-x) = y-x$, implying $-Rx = y-x$ or $(1-R)x = y$. ∎

It is to be noted that any G in $\mathcal{E}_\mathcal{K}$ is of the form

$$G_t = W(x\chi_{t]})\Gamma_t(Q)W(-x\chi_{t]})$$

with $Q = Q^* = Q^2, x \in$ range $(I-Q)$. In particular, for CCR flow of index 1 (i.e., when $\mathcal{K} = \mathbb{C}$), elements in $\mathcal{E}_\mathbb{C}$ are given by

$$\{G(I,0,0)\} \bigcup \{G(0,x,|x|^2) : x \in \mathbb{C}\}.$$

In other words projection cocycles are given by $G_t \equiv I$,

$$G_t = e^{-\frac{t\|x\|^2}{2}} |e(x\chi_{t]})\rangle\langle e(x\chi_{t]})| \otimes I_{[t}, \quad x \in \mathbb{C}.$$

This partially ordered set looks like $\mathbb{C} \cup \{\infty\}$ (no two elements of $\mathbb{C}$ are comparable, but every element in $\mathbb{C}$ is dominated by ∞). Every projection cocyle here (other than the identity) comes from a unit, i.e. $G_t = V_t V_t^*$, where $V = \{V_t : t \geq 0\}$ is a unit. Our understanding of units and projection cocycles of CCR flows yields the following result.

Corollary 7.7: Let θ be a primary, type I dilation of a unital quantum dynamical semigroup τ. Suppose the associated compression map maps normalized units $\mathcal{U}_0(\theta)$ to $\mathcal{U}_0(\tau)$ injectively. Then θ is the minimal dilation of τ.

Proof: If θ is elementary, primary dilation then it is automatically minimal (see Lemma 8.3). Otherwise, $\theta_t(\cdot) = U_t\gamma_t(\cdot)U_t^*$ for a CCR flow γ of $\Gamma(L^2(\mathbb{R}_+,\mathcal{K}))$ (for some Hilbert space $\mathcal{K}$), and a unitary cocycle U of γ. Now v_t is a unit of γ iff U_tv_t is a unit of θ and E_t is a projection cocycle of γ iff $U_tE_tU_t^*$ is a projection cocyle of θ.

Suppose $E = G(Q,x,||x||^2)$ is a projection cocycle of γ as in Theorem 7.6. Then for unit

$$v_t = e^{-pt}e(y\chi_{t]}), \quad (p,y) \in \mathbb{C} \times \mathcal{K}$$

of γ, we have

$$E_tv_t = e^{-(p+||x||^2-\langle x,y\rangle)t}e(x\chi_{t]} + Qy\chi_{t]}).$$

It is not hard to see that the map

$$(p,y) \mapsto (p + ||x||^2 - \langle x,y\rangle, x + Qy)$$

on $\mathbb{C} \times \mathcal{K}$ is injective iff $(Q,x,||x||^2) = (I,0,0)$. Then the result follows from remarks made just after Theorem 6.4. ∎

Powers [Po3] has proved that e_0-semigroups dominated by an E_0-semigroup forms a complete lattice on introducing a 'zero' element which is dominated by all e_0-semigroups. We can re-confirm this for CCR flows and hence for all type I flows as we have projection cocycles explicitly. We leave it to the reader to verify that if $E = G(R_1,y_1,||y_1||^2), F = G(R_2,y_2,||y_2||^2)$ are two projection cocycles of CCR flow γ then their meet $E \wedge F$ is given by $G(R_1 \wedge R_2, x, ||x||^2)$ if there exists x in range $(1 - R_1 \wedge R_2)$, such that $(1 - R_1)x = y_1$, $(1 - R_2)x = y_2$, it is 'zero' otherwise. Similarly join $E \vee F$ is given by $G(1 - [(1 - R_1 \vee R_2) \wedge S], z, ||z||^2)$, where S is the projection onto $[\mathbb{C}(y_1 - y_2)]^\perp$ and $z = [1 - R_1 \vee R_2) \wedge S]y_1 = [1 - R_1 \vee R_2) \wedge S]y_2$.

Now we proceed to study unitary local cocycles of CCR flows. It is clear that every unitary local cocycle of an E_0-semigroup gives rise to an automorphism of the associated product system. W. Arveson ([Ar1], Setion 8) has identified the automorphism group of CCR flow γ as $\mathcal{G} = \mathcal{U}(\mathcal{K}) \times \mathcal{K} \times \mathbb{R}$, where $\mathcal{U}(\mathcal{K})$ is the unitary group of $\mathcal{K}$ and the group law on $\mathcal{G}$ is given by

$$(U_1, u_1, c_1) \circ (U_2, u_2, c_2) = (U_1 U_2, u_1 + U_1 u_2, c_1 + c_2 + Im\langle u_1, U_1 u_2\rangle)$$

for $(U_1, u_1, c_1), (U_2, u_2, c_2) \in \mathcal{G}$. This group is well-known as the Heisenberg motion group. See [Th] for more information on this group.

For $(U, u, c) \in \mathcal{G}$, consider the family $\{g_t(U, u, c) : t \geq 0\}$ of unitary operators on $\mathcal{H}$, given by

$$g_t(U, u, c) = e^{-ict} W_t(u, U), \quad t \geq 0. \tag{7.5}$$

Here $W_t(u, U)$ are familiar (ampliated) Weyl operators [Pa], defined by

$$W_t(u, U)e(v) = e^{\frac{-t\|u\|^2}{2} - \langle u\chi_{t]}, Uv\rangle} e(u\chi_{t]} + Uv\chi_{t]} + v\chi_{[t}) \quad v \in L^2(\mathbb{R}_+, \mathcal{K}) \tag{7.6}$$

A routine computation shows that $\{g_t(U, u, c) : t \geq 0\}$ forms a unitary, local cocycle of γ and

$$g_t(U_1, u_1, c_1) \cdot g_t(U_2, u_2, c_2) = g_t((U_1, u_1, c_1) \circ (U_2, u_2, c_2)).$$

A little closer look at Arveson's work (essentially a simply application of the notion of 'divisible product systems') reveals that in case of CCR flows there is 1-1 correspondence between the automorphism group of the product system and the group of unitary local cocycles. We conclude that all unitary local cocycles of γ are given by (7.5).

We also note that conjugation by unitary local cocycles is an automorphism of $\mathcal{D}_\gamma$. So $G_t \mapsto g_t(U, u, c)^* G_t g_t(U, u, c)$ is an automorphism of the partially ordered set $\mathcal{D}_\mathcal{K}$. A straight-forward computation shows that this action of Heisenberg motion group on $\mathcal{D}_\mathcal{K}$ is given by

$$g_{(U,u,c)}((A,x,q)) = (U^*AU, U^*(x-(I-A)u), q-\langle x,u\rangle-\langle u,x\rangle+\langle u,(I-A)u\rangle). \tag{7.7}$$

If we look at matrices $[A,x,q]$ this action is particularly clear, as we have

$$1_\mathcal{K} - [g_{(U,u,c)}(A,x,q)] = \begin{bmatrix} 1 & u^* \\ 0 & U^* \end{bmatrix} \begin{bmatrix} q & -x^* \\ -x & 1-A \end{bmatrix} \begin{bmatrix} 1 & 0 \\ u & U \end{bmatrix}$$

$$= \begin{bmatrix} 1 & u^* \\ 0 & U^* \end{bmatrix} (1_\mathcal{K} - [A,x,q]) \begin{bmatrix} 1 & 0 \\ u & U \end{bmatrix}.$$

Note that the right-hand side is actually independent of c. In other words elements of the center of $\mathcal{G}$ act trivially on $\mathcal{D}_\mathcal{K}$. $\mathcal{D}_\mathcal{K}$ has much more structure than being just a partially ordered set and it is only natural that $g_{(U,u,c)}$ preserves most of them. For example we would like to point out that $g_{(U,u,c)}$ leaves $\mathcal{E}_\mathcal{K}$ invariant (as it should!) and in particular

$$g_{(U,x,0)}(Q,x,\|x\|^2) = (U^*QU, 0, 0). \tag{7.8}$$

Note that $g_t(U,x,0) = W_t(x,U)$ and the positive cocycle corresponding to $(U^*QU,0,0)$ is just $\Gamma_t(U^*QU)$. The last observation has an important role to play in our proof of the factorization theorem (Theorem 8.9).

Suppose ψ is an e_0-semigroup and is dominated by a type I flow θ. Say for example, γ is a CCR-flow and $\theta_t(\cdot) = U_t\gamma(\cdot)U_t^*$, and $\psi_t(\cdot) = U_t\gamma(\cdot)E_tU_t^*$ for a unitary cocycle U and projection cocycle E of γ. Then if $E = G(Q,x,\|x\|^2)$, we call rank $(I-Q)$ as the *deficiency index* of ψ in θ. Using the description of unitary, local cocyles of CCR flows it is easily verified that deficiency index does not depend upon the choice of unitary cocycle $\{U_t\}$ and it is a cocycle conjugacy invariant for the pair (ψ, θ).

Finally we record that if $K = G(A, x, q), H = G(B, y, r)$ are two positive, contractive, local cocycles of CCR flow γ then $R_t = H_t K_t H_t$ is the positive, contractive, local cocycle given by

$$R = G(BAB, y + Bx + BAy, q + 2r + \langle x, y \rangle + \langle y, x \rangle + \langle y, Ay \rangle). \qquad (7.9)$$

This computation is of some importance in view of Theorem 5.4.

8   Factorization theorem

Here we analyze general (not necessarily minimal) primary dilations of quantum dynamical semigroups. The objective is to show that, under suitable assumptions, any such dilation factorizes as minimal dilation tensored with some other E_0-semigroup.

We need the following definition of Powers. Recall that E_0-semigroups having units are called spatial. So they are of type I or II.

Definition 8.1: A spatial E_0-semigroup θ of $\mathcal{B}(\mathcal{H})$ is said to be in *standard form* if there exists a pure, normal state ω of $\mathcal{B}(\mathcal{H})$ such that $\rho \circ \theta_t(X) \to \omega(X)$, for $X \in \mathcal{B}(\mathcal{H})$, as $t \to \infty$, for every normal state ρ of $\mathcal{B}(\mathcal{H})$.

Let θ be a spatial E_0-semigroup in standard form. A state ω as in Definition 8.1 is said to be *absorbing*. It is clear that an absorbing state if it exists it is uniquly determined and that it is invariant, that is, $\omega \circ \theta_t = \omega, \forall t \geq 0$. As ω is a pure, normal state of $\mathcal{B}(\mathcal{H})$, we can find a unit vector a of $\mathcal{H}$, such that $\omega(X) = \langle a, Xa \rangle, X \in \mathcal{B}(\mathcal{H})$. Take $\mathcal{H}_0$ as the one-dimensional space spanned by a. Then P, the projection onto $\mathcal{H}_0$, is $|a\rangle\langle a|$. Invariance of ω shows that $\theta_t(P) \geq P, \quad \forall t \geq 0$. If $\theta_t(P)$ does not increase to 1, we can obtain a vector $v \in \mathcal{H}$, such that $\langle v, \theta_t(|a\rangle\langle a|)v \rangle = 0 \quad \forall t$. This is a contradiction as, $\langle v, \theta_t(|a\rangle\langle a|)v \rangle$ must converge to $\langle a, (|a\rangle\langle a|)a \rangle = 1$, by absorbing nature of ω. Hence $\theta_t(P)$ must increase to 1. We conclude that θ is a primary dilation of the trivial quantum dynamical semigroup $\gamma_t^0(X) \equiv X$ of $\mathcal{B}(\mathcal{H}_0)$.

We also note that $v(t) \equiv a$, is a unit of θ (in the vector picture). So existence of a pure, normal absorbing state actually forces θ to be spatial. Note that CCR flows are type I flows in standard form (vacuum vector gives rise to an absorbing state). We also consider the trivial semigroup $\gamma^{(0)}$ on $\mathcal{B}(\mathbb{C})$, defined by $\gamma_t^{(0)}(X) \equiv X$, as a type I_0 semigroup in standard form. Note that other type I_0 semigroups are never in standard form.

R.T. Powers has shown that any spatial E_0-semigroup θ, which is not a semigroup of automorphisms, is cocycle conjugate to an E_0-semigroup in standard form. Furthermore, if θ is of type I then the standard form is uniquely determined, that is, upto conjugacy there is unique E_0-semigroup in standard form to which θ is cocycle conjugate [Po3]. He also conjectures that the same result should be true for type II, E_0-semigroups as well. Aotani [Ao] has a different proof of Powers result. We prove uniqueness of standard form for a class which includes already known type I case, and perhaps contains some type II E_0-semigroups as well.

Definition 8.2: A spatial E_0-semigroup θ is said to be *amenable* if given any two isometric units $\{A_t\}, \{B_t\}$ of θ there exists a unitary local cocycle $\{U_t\}$ of θ such that $U_t A_t = B_t$ for all t.

From Arveson's computation of units and unitary, local cocyles of CCR flows we know that CCR flows are amenable. Clearly amenability is invariant under cocycle conjugacy. If an E_0-semigroup has index zero then upto scaling by an exponential factor it has unique unit. Hence, index zero E_0-semigroups are trivially amenable. So far it is not known as to whether there are type II_0 E_0-semigroups. As units of tensor products of E_0-semigroups are tensor products of units (see [Ar2]), we see that amenability is closed under taking tensor products. It is not known whether all spatial E_0-semigroups are amenable. We will show uniqueness of Powers' standard form for non-elementary amenable E_0-semigroups.

Lemma 8.3: Let θ be a primary dilation of a quantum dynamical semigroup τ. If θ is elementary then τ is elementary and moreover θ is the unique minimal dilation of τ.

Proof: As θ is an E_0-semigroup, it is elementary means that it is a semigroup of automorphisms. Say,

$$\theta_t(Z) = V_t Z V_t^*, \quad Z \in \mathcal{B}(\mathcal{H}), \quad t \geq 0$$

for some semigroup of unitaries $\{V_t : t \geq 0\}$ of $\mathcal{H}$. Then

$$\tau_t(X) = A_t X A_t^*, \quad X \in \mathcal{B}(\mathcal{H})_0, \quad t \geq 0$$

where A_t is the compression $P V_t P$ of V_t to $\mathcal{H}_0$. Clearly range $\theta_t(P) = V_t(\mathcal{H}_0)$. Moreover as θ is primary, $\theta_t(P) \uparrow 1$, giving us,

$$\overline{\mathrm{span}}\, \{V_t u : \quad u \in \mathcal{H}_0, \quad 0 \leq t < \infty\} = \mathcal{H}.$$

In other words $\{V_t : t \geq 0\}$ is the unique, minimal isometric dilation (in the sense of Sz. Nagy [SzF]) of semigroup of contractions $\{A_t : t \geq 0\}$. It follows that θ is the minimal dilation of τ. (see [Bh1], Theorem 2.6 for details).

Theorem 8.4 : Let θ, θ' be amenable, primary dilations of an elementary quantum dynamical semigroup τ. If θ, θ' are cocycle conjugate then they are conjugate as dilations of τ.

Proof : By replacing τ by its minimal dilation $\hat{\tau}$, if necessary, we can assume that τ is a semigroup of automorphisms. If θ is elementary then so is θ' as it is cocycle conjugate to θ. Then from Lemma 8.3, $\theta = \theta' = \tau$.

If θ, θ' are not elementary, it is clear that $\mathcal{H}_0^{\perp}$ is infinite dimensional for both θ, θ'. Then without loss of generality we can assume that both θ, θ' are acting on same $\mathcal{B}(\mathcal{H})$, with $\mathcal{H}_0 \subseteq \mathcal{H}$. Let $\tilde{\tau}, \tilde{\tau}'$ be respective induced semigroups for dilations θ, θ'. As τ is a semigroup of automorphisms the minimal dilation space $\hat{\mathcal{H}}$ is same as $\mathcal{H}_0$, for both θ, θ'. So we have, $P = \tilde{\tau}_t(P) = \tilde{\tau}_t'(P)$, where P is the projection on to $\mathcal{H}_0$. As θ' is cocycle conjugate to θ, we have a unitary cocycle $\{U_t : t \geq 0\}$ of θ such that

$$\theta_t'(Z) = U_t \theta_t(Z) U_t^* \quad Z \in \mathcal{B}(\mathcal{H}), t \geq 0.$$

From Proposition 3.5 (i) and its proof we know that

$$\tilde{\tau}_t(Z) = A_t Z A_t^*, \quad \tilde{\tau}_t'(Z) = A_t' Z (A_t')^* \quad Z \in \mathcal{B}(\mathcal{H})$$

for some isometric units $\{A_t\}$, $\{A_t'\}$ of θ, θ' respectively. Clearly $\{U_t^* A_t'\}$ is an isometric unit of θ. By amenability of θ there exists a unitary local cocycle R_t of θ such that $R_t A_t = U_t^* A_t'$ for all t. Now

$$\tilde{\tau}_t'(Z) = A_t' Z (A_t')^* = U_t R_t A_t Z A_t^* R_t^* U_t^* = U_t R_t \tilde{\tau}_t(Z) R_t^* U_t^*.$$

So replacing U_t by $U_t R_t$ we have $\theta_t'(Z) = U_t \theta_t(Z) U_t^*$ as well as $\tilde{\tau}_t'(Z) = U_t \tilde{\tau}_t(Z) U_t^*$ for $Z \in \mathcal{B}(\mathcal{H}), t \geq 0$. Now taking $Z = P$, we obtain $U_t P U_t^* = P$, that is U_t, U_t^* leave $\mathcal{H}_0$ invariant for all t. So the compression $V_t = P U_t P$, of U_t to $\mathcal{H}_0$ is unitary for every t. Note that as θ, θ' are dilations of τ,

$$\tau_t(X) = V_t \tau_t(X) V_t^* \quad \text{for } X \in \mathcal{B}(\mathcal{H}_0), t \geq 0.$$

Therefore,

$$\tau_t(X) V_t = V_t \tau_t(X) \quad \text{for } X \in \mathcal{B}(\mathcal{H}_0), t \geq 0.$$

But τ_t is an automorphism, hence V_t is a scalar for every t. Then strong continuity of $\{V_t : t \geq 0\}$ implies that $V_t = e^{itc}$, for some $c \in \mathbb{R}$. Replacing U_t by $U_t e^{-itc}$, we observe that it fixes every vector of $\mathcal{H}_0$, in other words θ, θ' are cocycle conjugate as dilations of τ. Now the result follows from Theorem 2.5. ∎

Corollary 8.5 : Two amenable E_0-semigroups in standard form are conjugate iff they are cocycle conjugate.

Proof : Any amenable E_0-semigroup in standard form is a primary dilation of the trivial unital quantum dynamical semigroup γ^0 of $\mathcal{B}(\mathbb{C})$, defined by $\gamma_t^0(X) \equiv X, X \in \mathcal{B}(\mathbb{C})$. So the result follows from Theorem 8.4. ∎

Suppose θ is a dilation of a quantum dynamical semigroup τ and γ is a spatial E_0-semigroup in standard form. Then $\theta \otimes \gamma$ can be considered as a dilation of τ, where now the initial space is $\mathcal{H}_0 \otimes a$, a being a unit vector such that the absorbing state ω of γ is given by $\omega(X) = \langle a, Xa \rangle$. Note that if θ is primary then so is $\theta \otimes \gamma$.

Theorem 8.6 (Factorization theorem for amenable dilations of elementary semigroups) : Let θ be an amenable primary dilation of an elementary quantum quantum dynamical semigroup τ. Then θ is conjugate as dilation to $\hat{\tau} \otimes \gamma$, where $\hat{\tau}$ is the unique minimal dilation of τ, and γ is a spatial E_0-semigroup in standard form. Furthermore, γ is uniquely determined upto conjugacy.

Proof : By replacing τ by $\hat{\tau}$ if necessary we can assume that τ is a semigroup of automorphisms. If θ is a semigroup of automorphisms then it is same as τ (Lemma 8.3). So we can take γ as $\gamma^{(0)}$, the trivial type I_0 semigroup of $\mathcal{B}(\mathbb{C})$.

Note that θ is spatial. Assume that θ is non-elementary. Let γ be an E_0-semigroup in standard form, cocycle conjugate to θ. (R.T. Powers [Po3], ensures for us the existence of such a γ). Consider the dilation $\hat{\tau} \otimes \gamma$. By Proposition 2.6 $\hat{\tau} \otimes \gamma$ is cocycle conjugate to γ. So it is cocycle conjugate to θ. Then Theorem 8.4 shows that θ and $\hat{\tau} \otimes \gamma$ are conjugate as dilations of τ. This proves the first part. Now suppose γ' is another E_0-semigroup in standard form such that $\hat{\tau} \otimes \gamma'$ is conjugate to θ. Clearly γ' is cocycle conjugate to γ. Both are in standard form. Hence they are conjugate to each other by Corollary 8.5.   ∎

As an immediate consequence of computing cocycles of CCR flows we have the following theorem, which shows the somewhat atomic nature of type I flows of index 1.

Theorem 8.7 : Let τ be a non-elementary unital quantum dynamical semigroup and let θ be a primary dilation of τ. Suppose θ is cocycle conjugate to CCR flow of index 1 then it is the unique minimal dilation of τ.

Proof : Let γ be a CCR flow of index 1. By looking at projection cocycles we see that any e_0-semigroup dominated by γ is either γ it self or is elementary. As θ is cocycle conjugate to γ, the same statement holds for θ. Now if the induced semigroup $\tilde{\tau}$ is equal to θ, as θ is primary we have minimality (see Remark 3.4). Otherwise, $\tilde{\tau}$ is elementary and we have a contradiction as compressions of elementary semigroups are elementary.   ■

Theorem 8.7 has ready applications. Many quantum dynamical semigroups with unbounded generators are known to have dilations cocycle conjugate to CCR flow of index 1. See for example ([Mo], [MS], [Fa2], [Jo]). These dilations have been constructed by various methods. Now Theorem 8.7 says that all such dilations are minimal. Minimality comes for free. Note then that the method of construction becomes irrelevant as there is a unique minimal dilation up to unitary equivalence. We also get almost complete information about quantum dynamical semigroups dominated by the given semigroup.

Theorem 8.8: Suppose a unital quantum dynamical semigroup τ has a type I dilation then the minimal dilation of τ is also of type I, of equal or lower index.

Proof: To begin with as minimal dilations of unital elementary semigroups are type I flows of index zero (i.e., semigroups of automorphisms) we can assume that τ is not elementary. Let θ be a type I dilation of τ. If index $(\theta) = 0$, then θ and hence τ are elementary. So we can assume index $(\theta) \geq 1$. Now we have $\theta_t(\cdot) = U_t \gamma_t(\cdot) U_t^*$, where $\gamma = \gamma^{\mathcal{K}}$, is a CCR flow with $\dim(\mathcal{K}) =$ index (θ), and $\{U_t\}$ is a unitary cocycle of γ. Hence $\tilde{\tau}(\cdot) = U_t \gamma_t(\cdot) E_t U_t^*$, for some projection cocycle $\{E_t\}$ of γ. By replacing U_t by $U_t W_t(u, U)$, for suitable $u \in \mathcal{K}$, unitary $U \in \mathcal{B}(\mathcal{K})$ (see (7.8) and remarks on it), we can assume that $E_t = \Gamma_t(Q)$, for some projection Q of $\mathcal{K}$.

First assume $0 < Q < I$. Decomposing $\mathcal{K}$ as $\mathcal{K} = \mathcal{K}_0 \oplus \mathcal{K}_1$, where $\mathcal{K}_0 =$

range $Q, \mathcal{K}_1 = (\text{range } Q)^\perp$, we can identify $\mathcal{H} = \Gamma(L^2(\mathbb{R}_+, \mathcal{K}))$ with $\mathcal{H}^0 \otimes \mathcal{H}^1$ where $\mathcal{H}^0 = \Gamma(L^2(\mathbb{R}_+, \mathcal{K}_0))$ and $\mathcal{H}^1 = \Gamma(L^2(\mathbb{R}_+, \mathcal{K}_1))$, so that $e(v) = e(Qv) \otimes e((1-Q)v)$ for $v \in L^2(\mathbb{R}_+, \mathcal{K})$. Then, on taking CCR flows $\gamma^0 = \gamma^{\mathcal{K}_0}$, $\gamma^1 = \gamma^{\mathcal{K}_1}$, $\gamma_t = \gamma_t^0 \otimes \gamma_t^1$, consider the e_0-semigroup $\tilde{\gamma}_t = (\gamma_t^0 \otimes \gamma_t^1) E_t$, where $E_t = \Gamma_t(Q)$ is given by

$$E_t(e(u) \otimes e(v)) = e(u) \otimes e(v \chi_{[t})$$

for $u \in L^2(\mathbb{R}_+, \mathcal{K}_0), v \in L^2(\mathbb{R}_+, \mathcal{K}_1)$. Now looking at the isometries $\{A_t : t \geq 0\}$ defined by

$$A_t(e(u) \otimes e(v)) = e(u) \otimes (S_t v)$$

where $\{S_t : t \geq 0\}$ are shifts as in Section 7, it is clear that

$$\left[\gamma_t^0(X) \otimes \gamma_t^1(Y)\right] E_t = A_t \left[\gamma_t^0(X) \otimes Y\right] A_t^*.$$

In other words $\tilde{\gamma}_t$ and $\gamma_t^0 \otimes id$ are cocycle conjugate through isometries. Recalling the definition of $\bar{\tau}$ and (3.5) from the proof of Proposition 3.5 (ii) we have equalities

$$\begin{aligned}
\tilde{\gamma}_t(Z) &= A_t(\gamma_t^0 \otimes id)(Z) A_t^*, \\
\tilde{\tau}_t(Z) &= U_t \tilde{\gamma}_t(Z) U_t^*, \\
\tilde{\tau}_t(Z) &= V_t \bar{\tau}_t(Z) V_t^*, \quad Z \in \mathcal{B}(\mathcal{H}), \ t \geq 0,
\end{aligned}$$

where A_t, V_t are isometric and U_t is unitary for every t. Combining the three relations gives

$$\bar{\tau}_t(Z) = B_t(\gamma_t^0 \otimes id)(Z) B_t^*, \quad Z \in \mathcal{B}(\mathcal{H}), \ t \geq 0,$$

where $B_t = V_t^* U_t A_t$. Clearly $B_t B_t^* = 1$, as $\bar{\tau}_t$ is unital. Furthermore, $B_t^* B_t = A_t^* U_t^* V_t V_t^* U_t A_t = A_t^* U_t^* \tilde{\tau}_t(1) U_t A_t = A_t^* \tilde{\gamma}_t(1) A_t = (\gamma_t^0 \otimes id)(1) = 1$. A routine computation shows that $\{B_t\}$ is a cocycle of $(\gamma_t^0 \otimes id)$ (continuity follows from measurability, see Appendix A). Now as $\bar{\tau}$ is conjugate to $\hat{\tau}$, and $\gamma_t^0 \otimes id$ is cocycle conjugate to γ_t^0 (Proposition 2.6), we obtain that $\hat{\tau}$ is cocycle conjugate to the CCR flow γ_t^0.

If $Q = I$, then $\tilde{\tau} = \theta$, and as $\hat{\tau}$ and $\tilde{\tau}$ are cocycle conjugate (Proposition 3.5 (ii)), the required result follows. If $Q = 0$, then $\tilde{\tau}$ is elementary, and hence τ is also elementary. We have already excluded this possibility. ∎

Now we come to the main result of this section.

Theorem 8.9 (Factorization theorem for type I dilations) : Let θ be a primary type I dilation of a unital quantum dynamical semigroup τ. Then either θ is the minimal dilation $\hat{\tau}$ of τ or it is conjugate as dilation to $\hat{\tau} \otimes \gamma$, where γ is a CCR flow. The index of γ in this factorization is uniquely determined by θ and τ.

Proof: *Case (i),* index $(\theta) = 0$: Here θ is semigroup of *-automorphisms. From Lemma 8.3 θ is the unique minimal dilation of τ.

Case (ii), index $(\theta) \geq 1$: Let $\mathcal{K}$ be a Hilbert space with $\dim(\mathcal{K}) = $ index (θ). Now θ is cocycle conjugate to CCR flow γ acting on $\mathcal{B}(\Gamma(L^2(\mathbb{R}_+, \mathcal{K})))$. By replacing (θ, τ) by conjugate (conjugate as dilation) semigroups we can assume that θ, τ are acting on $\mathcal{B}(\mathcal{H}), \mathcal{B}(\mathcal{H}_0)$, where $\mathcal{H} = \Gamma(L^2(\mathbb{R}_+, \mathcal{K})), \mathcal{H}_0$ is a subspace of $\mathcal{H}$, with a unitary cocycle $\{U_t : t \geq 0\}$ of γ such that

$$\theta_t(Z) \;=\; U_t \gamma_t(Z) U_t^* \quad Z \in \mathcal{B}(\mathcal{H}), t \geq 0. \tag{8.1}$$

As in the proof of Theorem 8.8, by changing the unitary cocycle $\{U_t\}$, if needed, we have the induced semigroup

$$\tilde{\tau}_t(Z) \;=\; U_t\gamma_t(Z)E_tU_t^* \quad Z \in \mathcal{B}(\mathcal{H}), t \geq 0, \tag{8.2}$$

with projection cocycle $E_t = \Gamma_t(Q)$, for some projection Q of $\mathcal{K}$.

Subcase (i), $Q = I$: Now θ is the minimal dilation of τ and there is nothing to prove.

Subcase (ii), $0 < Q < I$: Here again as in the proof of Theorem 8.8, we can decompose $\mathcal{H}$ as $\mathcal{H} = \mathcal{H}^0 \otimes \mathcal{H}^1$, where $\mathcal{H}^0 = \Gamma(L^2(\mathbb{R}_+, \mathcal{K}_0)), \mathcal{H}^1 = \Gamma(L^2(\mathbb{R}_+, \mathcal{K}_1))$ with

$$\gamma_t = \gamma_t^0 \otimes \gamma_t^1, \tag{8.3}$$

$$E_t(e(u) \otimes e(v)) = e(u) \otimes e(v\chi_{[t}) \tag{8.4}$$

for $u \in L^2(\mathbb{R}_+, \mathcal{K}_0), v \in L^2(\mathbb{R}_+, \mathcal{K}_1)$. The initial space $\mathcal{H}_0$ and the minimal dilation space $\hat{\mathcal{H}}$ are some subspaces of $\mathcal{H}$. As $\tilde{\tau}$ is not elementary (Clear from (8.2)) τ is also not elementary, and also as $0 < Q < I, \theta$ is not minimal. Then it is clear that $\hat{\mathcal{H}}, \hat{\mathcal{H}}^\perp$ are infinite dimensional. So we can obtain a unitary $M : \mathcal{H} \rightarrow \mathcal{H}$ such that $M(\hat{\mathcal{H}}) = \mathcal{H}^0 \otimes e(0)$. Now instead of θ, τ consider the conjugate semigroups $\bar{\theta}, \bar{\tau}$ given by

$$\bar{\theta}_t(Z) = M\theta_t(M^*ZM)M^* \quad \text{for } Z \in \mathcal{B}(\mathcal{H}), t \geq 0,$$

$$\bar{\tau}_t(X) = M\tau_t(M^*XM)M^* \quad \text{for } X \in \mathcal{B}(\bar{\mathcal{H}}_0), t \geq 0,$$

where the new initial space $\bar{\mathcal{H}}_0$ is given by $\bar{\mathcal{H}}_0 = M(\mathcal{H}_0) \subseteq \mathcal{H}^0 \otimes e(0)$. We note that

$$\bar{\theta}_t(Z) \;=\; M\theta_t(M^*ZM)M^*$$

$$= M\theta_t(M^*)U_t\gamma_t(Z)U_t^*\theta_t(M)M^*$$

$$= V_t\gamma_t(Z)V_t^*,$$

where $V_t = M\theta_t(M^*)U_t$. Clearly $\{V_t : t \geq 0\}$ is a unitary cocycle of γ. Further, the induced semigroup $\tilde{\bar{\tau}}$ of $\bar{\tau}$ is given by

$$\tilde{\bar{\tau}}_t(Z) = M\tilde{\tau}_t(M^*ZM)M^*$$

$$= MU_t\gamma_t(M^*ZM)E_tU_t^*M^*$$

$$= M\theta_t(M^*)U_t\gamma_t(Z)E_tU_t^*\theta_t(M)M^*$$

$$= V_t\gamma_t(Z)E_tV_t^*, \quad Z \in \mathcal{B}(\mathcal{H}), t \geq 0.$$

We can replace the pair (θ, τ) by the pair $(\bar{\theta}, \bar{\tau})$. This means that with no loss of generality we can assume that the initial space $\mathcal{H}_0$ is a subspace of $\mathcal{H}^0 \otimes e(0)$, the minimal dilation space $\hat{\mathcal{H}}$ is $\mathcal{H}^0 \otimes e(0)$ and at the same time $\theta, \tilde{\tau}, E$ are given by (8.1) - (8.4). Once this is done we claim that U_t, U_t^* leave $\hat{\mathcal{H}} = \mathcal{H}^0 \otimes e(0)$ invariant for every t. Let R^0 be the projection onto $\mathcal{H}^0 \otimes e(0)$. We have $\tilde{\tau}_t(R^0) = R^0$, as R^0 is the projection onto the minimal dilation space. Hence

$$U_t(\gamma_t^0 \otimes \gamma_t^1)(R^0)E_tU_t^* = R^0.$$

But clearly the explicit form (8.4) of E_t gives

$$(\gamma_t^0 \otimes \gamma_t^1)(R^0)E_t = R^0$$

for all t. Hence $U_tR^0U_t^* = R^0$, or $U_tR^0 = R^0U_t, U_t^*R^0 = R^0U_t^*$ for every t, proving our claim. Let D_t be the compression of U_t to $\mathcal{H}^0 \otimes e(0)$, thought of as an operator on $\mathcal{H}^0$. (We identify $\hat{\mathcal{H}} = \mathcal{H}^0 \otimes e(0)$ with $\mathcal{H}^0$). As U_t, U_t^* leave $\mathcal{H}^0 \otimes e(0)$ invariant D_t is a unitary for every t. Moreover $\{D_t : t \geq 0\}$ is a unitary cocycle of γ_t^0. Now the minimal dilation $\hat{\tau}$ of τ is got by compressing θ to $\mathcal{B}(\hat{\mathcal{H}})$. But for $Y \in \mathcal{B}(\mathcal{H}^0)$.

$$
\begin{aligned}
R^0\theta_t(Y \otimes |e(0)\rangle\langle e(0)|)R^0 &= R^0 U_t[(\gamma_t^0 \otimes \gamma_t^1)(Y \otimes |e(0)\rangle\langle e(0)|)]U_t^* R^0 \\
&= R^0 U_t R^0[(\gamma_t^0 \otimes \gamma_t^1)(Y \otimes |e(0)\rangle\langle e(0)|)]R^0 U_t^* R^0 \\
&= D_t \gamma_t^0(Y) D_t^* \otimes |e(0)\rangle\langle e(0)|
\end{aligned}
$$

Then it is clear that the minimal dilation $\hat{\tau} = \{\hat{\tau}_t : t \geq 0\}$ is given by

$$
\hat{\tau}_t(Y) = D_t \gamma_t^0(Y) D_t^*, \quad Y \in \mathcal{B}(\mathcal{H}^0), t \geq 0.
$$

Now we re-write (8.1) as

$$
\theta_t(Z) = B_t[(\hat{\tau}_t \otimes \gamma_t^1)(Z)]B_t^*,
$$

where $B_t = U_t(D_t^* \otimes 1)$. Clearly $B = \{B_t : t \geq 0\}$ is a unitary cocycle of $\hat{\tau} \otimes \gamma^1$. Moreover it fixes $\mathcal{H}^0 \otimes e(0)$ as

$$
B_t(z \otimes e(0)) = U_t(D_t^* \otimes 1)(z \otimes e(0)) = U_t(D_t^* z \otimes e(0)) = D_t D_t^* z \otimes e(0) = z \otimes e(0)
$$

for $z \in \mathcal{H}^0$. Then from Theorem 2.5, θ and $\hat{\tau} \otimes \gamma^1$ are conjugate as dilations.

Subcase (iii), $Q = 0$: Here $\tilde{\tau}$ is elementary. Consequently its compressions $\tau, \hat{\tau}$ are also elementary. So the result follows from Theorem 8.6 as type I flows are amenable.

Uniqueness of the factorization is clear once we observe that index (γ) is just the deficiency index of $\tilde{\tau}$ in θ (see remarks towards the end of Section 7). ∎

Corollary 8.10: Let θ, θ' be two primary type I dilations of a quantum dynamical semigroup τ. Suppose that index θ = index $\theta' < \infty$. Then θ, θ' are conjugate as dilations of τ. In particulr, if θ is minimal so is θ'.

Proof : We have θ, θ' conjugate as dilations to $\hat{\tau} \otimes \gamma, \hat{\tau} \otimes \gamma'$ respectively for some type I flows γ, γ' in standard form. So only thing we need to show is that index $(\gamma) =$ index (γ'). But this is clear as we have the addition formula of Arveson for tensor products of E_0-semigroups ([Ar1], [Ar2]), giving us index $(\theta) =$ index $(\hat{\tau}) +$ index $(\gamma) =$ index $(\hat{\tau}) +$ index $(\gamma') =$ index (θ'). and index $\theta =$ index $\theta' < \infty$ by assumption.   ∎

Factorization theorem fails in the discrete case. See Appendix B for the correction needed.

9 Hudson-Parthasarathy cocycles

Let $\mathcal{H}_0, \mathcal{K}$ be complex separable Hilbert spaces. Then Hudson-Parthasarathy quantum stochastic diferential equations can be written on the space $\tilde{\mathcal{H}} = \mathcal{H}_0 \otimes \Gamma(L^2(I\!R_+, \mathcal{K}))$. Here w $\mathcal{H}_0$ (identified with $\mathcal{H}_0 \otimes e(0)$) is known as the initial space and $\mathcal{H} = \Gamma(L^2(I\!R_+, \mathcal{K}))$ as the noise space. On $\tilde{\mathcal{H}}$ we have the E_0-semigroup $\tilde{\gamma} = id \otimes \gamma$, where $\gamma = \gamma^{\mathcal{K}}$ is the CCR flow on $\mathcal{H}$. Note that by a result of Arveson (Proposition 2.6) we know that $\tilde{\gamma}$ is cocycle conjugate to $\gamma^{\mathcal{K}}$. So $\tilde{\gamma}$ is also a type I flow with same index as γ namely $\dim(\mathcal{K})$. Hudson-Parthasarathy theory provides us with many unitary cocycles of $\tilde{\gamma}$.

We adapt the necessary notation from [Pa] (page 221-222 and also 179-181). For every positive t we denote $\mathcal{H}_0 \otimes \Gamma(L^2[0,t], \mathcal{K})$ by $\tilde{\mathcal{H}}_{t]}$ and $\Gamma([t, \infty), \mathcal{K})$ by $\tilde{\mathcal{H}}_{[t}$. Then we have the natural decomposition of $\tilde{\mathcal{H}}$ as $\tilde{\mathcal{H}}_{t]} \otimes \tilde{\mathcal{H}}_{[t}$. Note that $\tilde{\mathcal{H}}_{0]}$ is same as $\mathcal{H}_0$ and in general $\tilde{\mathcal{H}}_{t]}$ is to be identified with $\tilde{\mathcal{H}}_{t]} \otimes e(0_{[t})$, $e(0_{[t})$ denoting the vacuum vector in $\tilde{\mathcal{H}}_{[t}$.

By a *bounded process* X on $\tilde{\mathcal{H}}$ we simply mean a family $X = \{X_t \in \mathcal{B}(\tilde{\mathcal{H}}) : t \geq 0\}$ of bounded operators. It is said to be *adapted* if $X_t = X_t^0 \otimes 1_{[t}$, for some $X_t^0 \in \mathcal{B}(\tilde{\mathcal{H}}_{t]})$, ($1_{[t}$ is the identity operator on $\mathcal{H}_{[t})$. Note that if X is local for $\tilde{\gamma}$ then X_t is of the form $1_{0]} \otimes Y_t \otimes 1_{[t}$ for some $Y_t \in \mathcal{B}(\Gamma(L^2([0,t], \mathcal{K})))$. In particular, local processes are adapted but not conversely.

Fix an orthonormal basis $\{e_k : k \geq 1\}$ for $\mathcal{K}$. For any u in $\mathcal{H}$, write

$$u_i(t) = \langle u(t), e_i \rangle, \quad u^i(t) = \langle e_i, u(t) \rangle \text{ if } i \geq 1,$$
$$u_0(t) = u^0(t) = 1;$$

and define

$$\mathcal{M} = \{u | u \in \mathcal{H}, u_i(t) = 0 \text{ for all but a finite number of indices } i.\}.$$

Note that if $\mathcal{K}$ is finite dimensional $\mathcal{M}$ is whole of $\mathcal{H}$.

Let τ be a unital quantum dynamical semigroup on $\mathcal{B}(\mathcal{H}_0)$ with bounded generator. Then by ([GKS], [Li], [EC]) we know that its generator has a very special form. Namely if we denote the generator by $\mathcal{L}$ so that $\tau_t(X) = e^{t\mathcal{L}}(X), X \in \mathcal{B}(\mathcal{H}_0), t \geq 0$, then there exists a family of bounded operators $\{L_k \in \mathcal{B}(\mathcal{H}_0) : k \geq 1\}$ and a self-adjoint operator $H \in \mathcal{B}(\mathcal{H}_0)$ such that $\sum_{k \geq 1} L_k^* L_k$ is strongly convergent, and

$$\mathcal{L}(X) = i[H, X] - \frac{1}{2} \sum_{k \geq 1} (L_k^* L_k X + X L_k^* L_k - 2 L_k^* X L_k) \qquad (9.1)$$

for every $X \in \mathcal{B}(\mathcal{H}_0)$. Now to obtain a dilation of τ using quantum stochastic calculus we fix one such representation of $\mathcal{L}$ and consider a Hilbert space $\mathcal{K}$ with $\dim(\mathcal{K})$ equal to number of L_k's. To avoid trivialities we assume that this number is non-zero.

Let $\{S_j^i : i, j \geq 1\}$ be bounded operators on $\mathcal{H}_0$ such that $\sum_{i,j \geq 1} S_j^i \otimes |e_i\rangle\langle e_j|$ is a unitary operator in $\mathcal{H}_0 \otimes \mathcal{K}$. Define

$$L_j^i = \begin{cases} S_j^i - \delta_{ij} & \text{if } i, j \geq 1; \\ L_i & \text{if } i \geq 1, j = 0; \\ -\sum_{k \geq 1} L_k^* S_j^k & \text{if } j \geq 1, i = 0; \\ -(iH + \frac{1}{2} \sum_{k \geq 1} L_k^* L_k) & \text{if } i = j = 0. \end{cases}$$

Then by Theorem 27.8 of [Pa] (adapted from [HP], [Mo]) there exists a unique unitary operator valued adapted process $U = \{U_t : t \geq 0\}$ on $\mathcal{B}(\tilde{\mathcal{H}})$

satisfying the quantum stochastic differential equation

$$dU = \left(\sum_{i,j \geq 0} L_j^i d\Lambda_i^j \right) U, \quad U_0 = I \tag{9.2}$$

on $\mathcal{M}$. Here $d\Lambda_i^j$ refer to differentials with respect to the fundamental processes of time, creation, conservation and annihilation. We will not go in to conceptual and technical details involved here. Instead we just infer from (9.2) that we have an adapted unitary right cocycle $U = \{U_t : t \geq 0\}$ of $\tilde{\gamma}$ such that

$$
\begin{aligned}
&\langle fe(u), U_t ge(v) \rangle \\
&= \langle fe(u), ge(v) \rangle + \int_0^t \left\{ \sum_{i,j \geq 0} u_i(s) v^j(s) \langle fe(u), (L_j^i \otimes I) U_s ge(v) \rangle \right\} ds
\end{aligned}
\tag{9.3}
$$

for $f, g \in \mathcal{H}_0, u, v \in \mathcal{M}$, where $fe(u), ge(v)$ are just short-hand notations for $f \otimes e(u)$, and $g \otimes e(v)$ in $\tilde{\mathcal{H}}$. Note that u, v are restricted to vary only in $\mathcal{M}$ simply to avoid possible convergence problems in summing up terms inside the integral (Now we are having effectively a finite sum). Note that as strong and weak topologies coincide on the set of unitaries, U is a right cocycle of unitaries if and only if $\{U_t^* : t \geq 0\}$ is a left cocyle of unitaries.

We then have an E_0-semigroup η of $\mathcal{B}(\tilde{\mathcal{H}})$ defined by

$$\eta_t(Z) = U_t^* \tilde{\gamma}_t(Z) U_t \qquad \text{for } Z \in \mathcal{B}(\tilde{\mathcal{H}}). \tag{9.4}$$

Conventionally the family $j = \{j_t : t \geq 0\}$ of representations of $\mathcal{B}(\mathcal{H}_0)$ defined by

$$j_t(X) = \eta_t(X \otimes 1_{\mathcal{H}}) = U_t^*(X \otimes 1_{\mathcal{H}}) U_t, \quad X \in \mathcal{B}(\mathcal{H}_0) \tag{9.5}$$

is known as the Evans-Hudson flow (or EH flow) associated with the Hudson-Parthasarathy cocycle U. By a small modification of the terminology we call the E_0-semigroup η as the Evans-Hudson flow.

As we identify $\mathcal{H}_0$ with $\mathcal{H}_0 \otimes e(0), X \in \mathcal{B}(\mathcal{H}_0)$ is to be identified with $X \otimes |e(0)\rangle\langle e(0)|$ in $\mathcal{B}(\tilde{\mathcal{H}})$. So for $X \in \mathcal{B}(\mathcal{H}_0)$, by $\eta_t(X)$ we mean $\eta_t(X \otimes |e(0)\rangle\langle e(0)|)$, which is not same as $j_t(X)$. However adaptedness of U_t gives

$$\langle fe(0), \eta_t(X)ge(0)\rangle = \langle fe(0), j_t(X)ge(0)\rangle \quad f, g \in \mathcal{H}_0, X \in \mathcal{B}(\mathcal{H}_0)$$

that is, compressions of $\eta_t(X), j_t(X)$ to $\mathcal{B}(\mathcal{H}_0)$ are same. Then by standard computation (see Corollary 27.9 of [Pa]) we deduce

$$\langle fe(0), \eta_t(X)ge(0)\rangle = \langle fe(0), Xge(0)\rangle + \int_0^t \langle fe(0), \eta_s(\mathcal{L}(X))ge(0)\rangle ds.$$

As $\mathcal{L}$ is the generator of τ, η is a dilation of τ. The problem is to determine the minimality of this dilation and also to compute the induced semigroup $\tilde{\tau}$, when η is not minimal.

Now any cocycle $K = \{K_t : t \geq 0\}$ of η is given by $K_t = U_t^*(1_{\mathcal{H}_0} \otimes G_t)U_t$, for some cocycle $G = \{G_t : t \geq 0\}$ of γ. Recall that in Section 7 we have parametrized positive, contractive, local cocycles of γ (or equivalently the set $\mathcal{D}_\gamma$) by $\{G(A, x, q) : (A, x, q) \in \mathcal{D}_\mathcal{K}\}$. Fix any $(A, x, q) \in \mathcal{D}_\mathcal{K}$ and consider the associated cocycle G of γ, defined by (7.1). We note that for any $X \in \mathcal{B}(\mathcal{H}_0), \{X \otimes G_t : t \geq 0\}$ is an adapted process satisfying

$$\langle fe(u), (X \otimes G_t)ge(v)\rangle$$
$$= \langle f, g\rangle e^{-qt + \langle x\chi_{t]}, v\rangle + \langle u, x\chi_{t]}\rangle + \langle u, (A-I)v\chi_{t]}\rangle + \langle u, v\rangle}$$
$$= \langle fe(u), (X \otimes I)ge(v)\rangle$$

$$+ \int_0^t \left\{ \sum_{i,j \geq 0} u_i(s) v^j(s) \langle fe(u), m_j^i(X \otimes G_s) ge(v) \rangle \right\} ds$$

for $f, g \in \mathcal{H}_0, u, v \in \mathcal{M}$, where

$$m_j^i = \begin{cases} \langle e_i, (A - I)e_j \rangle & \text{if } i, j \geq 1; \\ \langle e_i, x \rangle & \text{if } i \geq 1, j = 0; \\ \langle x, e_j \rangle & \text{if } j \geq 1, i = 0; \\ -q & \text{if } i = j = 0. \end{cases}$$

Then by quantum Ito's formula (See Corollary 27.2 of [Pa]) one can compute the integral equations satisfied by the process $\{(X \otimes G_t)U_t : t \geq 0\}$, and we have

$$\langle fe(u), (X \otimes G_t)U_t ge(v) \rangle$$

$$= \langle fe(u), (X \otimes I)ge(v) \rangle$$

$$+ \int_0^t \{ \sum_{i,j \geq 0} u_i(s) v^j(s) \langle fe(u), ([XL_j^i + \bar{m}_i^j X + \sum_{k \geq 1} \bar{m}_i^k X L_j^k] \otimes G_s)U_s ge(v) \rangle \} ds$$

on $\mathcal{H}_0 \otimes e(\mathcal{M})$. (Here $\bar{m}_i^j$ stands for complex conjugate of m_i^j.) Another such computation yields

$$\langle fe(u), U_t^*(X \otimes G_t)U_t ge(v) \rangle$$

$$= \langle fe(u), (X \otimes I)ge(v) \rangle$$

$$+ \int_0^t \{ \sum_{i,j \geq 0} u_i(s) v^j(s) \langle fe(u), U_s^*(\psi_j^i(X) \otimes G_s)U_s ge(v) \rangle \} ds \quad (9.6)$$

on $\mathcal{H}_0 \otimes e(\mathcal{M})$ where

$$\psi_j^i(X) = XL_j^i + \bar{m}_i^j X + \sum_{k \geq 1} \bar{m}_i^k X L_j^k + (L_i^j)^* X + \sum_{l \geq 1} (L_i^l)^* X L_j^l$$

$$+ \sum_{l \geq 1} (L_i^l)^* \bar{m}_l^j X + \sum_{l,k \geq 1} (L_i^l)^* \bar{m}_l^k X L_j^k.$$

In the language of quantum stochastic calculus, $h_t(X) = U_t^*(X \otimes G_t)U_t, X \in \mathcal{B}(\mathcal{H}_0)$ is an adapted process satisfying quantum stochastic differential equation

$$dh_t(X) = \sum_{i,j \geq 0} h_s(\psi_j^i(X))d\Lambda_i^j, \quad h_0(X) = X \otimes I. \tag{9.7}$$

Enlarge the family of bounded operators $\{L_i \in \mathcal{B}(\mathcal{H}_0) : i \geq 1\}$ to $\{L_i \in \mathcal{B}(\mathcal{H}_0) : i \geq 0\}$ by setting $L_0 = I$. We say that the sequence $\{L_i : i \geq 0\}$ is *linearly independent* if $\sum_{i \geq 0} c_i L_i = 0$ (convergence in strong operator topology), for some $c_0, c_1, \cdots$ in $\mathbb{C}$ such that $\sum |c_i|^2 < \infty$, then $c_i \equiv 0$ for all i.

Theorem 9.1: The Evans-Hudson flow $\eta = \{\eta_t : \eta_t(\cdot) = U_t^*(\tilde{\gamma}_t(\cdot))U_t, t \geq 0\}$ coming from the Hudson-Parthasarathy cocycle $\{U_t\}$ of (9.2) is a minimal dilation of τ if and only if $\{L_i : i \geq 0\}$ are linearly independent.

Proof: We know that an arbitrary quantum dynamical semigroup β dominated by η is of the form

$$\beta_t(Z) = \eta_t(Z)U_t^*(1_{\mathcal{H}_0} \otimes G_t)U_t \quad Z \in \mathcal{B}(\tilde{\mathcal{H}})$$

for some positive, contractive, local cocycle $G = G(A, x, q)$ of γ. For $X \in \mathcal{B}(\mathcal{H}_0), Y \in \mathcal{B}(\mathcal{H})$ we have

$$\begin{aligned}\beta_t(X \otimes Y) &= U_t^*(X \otimes \gamma_t(Y))U_t U_t^*(1_{\mathcal{H}_0} \otimes G_t)U_t \\ &= U_t^*(X \otimes \gamma_t(Y)G_t)U_t, \quad t \geq 0.\end{aligned}$$

Let α be the compression of β to $\mathcal{B}(\mathcal{H}_0)$. Then using adaptedness of U_t, G_t we have

$$\langle f, \alpha_t(X)g \rangle = \langle fe(0), \beta_t(X \otimes |e(0)\rangle\langle e(0)|)ge(0)\rangle$$

$$= \langle fe(0), U_t^*(X \otimes \gamma_t(|e(0)\rangle\langle e(0)|)G_t)U_t ge(0)\rangle$$

$$= \langle fe(0), U_t^*(X \otimes \gamma_t(1)G_t)U_t ge(0)\rangle$$

$$= \langle fe(0), U_t^*(X \otimes G_t)U_t ge(0)\rangle, \quad \text{for } f, g, \in \mathcal{H}_0, X \in \mathcal{B}(\mathcal{H}_0).$$

Setting $u = v = 0$ in (9.6) we pick up only the $i = j = 0$ term, that is,

$$\langle f, \alpha_t(X)g \rangle = \langle fe(0), (X \otimes I)ge(0)\rangle$$
$$+ \int_0^t \langle fe(0), U_s^*(\psi_0^0(X) \otimes G_s)U_s ge(0)\rangle ds$$
$$= \langle f, Xg \rangle + \int_0^t \langle f, \alpha_s(\psi_0^0(X))g \rangle ds.$$

So, if we denote the generator of α by $\mathcal{L}_G$, we have

$$\mathcal{L}_G(X) = \psi_0^0(X) = XL_0^0 + \bar{m}_0^0 X + \sum_{k \geq 1} \bar{m}_0^k X L_0^k + (L_0^0)^* X$$
$$+ \sum_{l \geq 1} (L_0^l)^* X L_0^l + \sum_{l \geq 1} (L_0^l)^* \bar{m}_l^0 X + \sum_{k,l \geq 1} (L_0^l)^* \bar{m}_l^k X L_0^k,$$

for $X \in \mathcal{B}(\mathcal{H}_0)$. Substituting the values of L_j^l's, M_j^l's we get

$$\mathcal{L}_G(X) = X(-iH - \frac{1}{2}\sum_{k \geq 1} L_k^* L_k) + (-q)X + \sum_{k \geq 1} \langle x, e_k \rangle X L_k$$
$$+ (iH - \frac{1}{2}\sum_{k \geq 1} L_k^* L_k)X + \sum_{k \geq 1} L_k^* X L_k + \sum_{k \geq 1} \langle e_k, x \rangle L_k^* X$$
$$+ \sum_{k,l \geq 1} \langle e_l, (A - I)e_k \rangle L_l^* X L_k$$

As we know that $x = (I - A)^{1/2}a$, $q = p + \|a\|^2$ for some $a \in \mathcal{K}$ and $p \geq 0$, we have

$$\mathcal{L}_G(X) = \mathcal{L}(X) - pX$$
$$-\sum_{n \geq 1} [\langle a, e_n \rangle I - \sum_{l \geq 1} \langle (I - A)^{1/2}e_l, e_n \rangle L_l^*] X [\langle e_n, a \rangle I - \sum_{k \geq 1} \langle e_n, (I - A)^{1/2}e_k \rangle L_k].$$

Note that $\mathcal{L} - \mathcal{L}_G$ is a completely positive map. If for some $(A, x, q) \in \mathcal{D}_{\mathcal{K}}, \mathcal{L}_G(X) \cong \mathcal{L}(X)$, then we have

$$p = 0, \quad \langle e_n, a \rangle I - \sum_{k \geq 1} \langle e_n, (I - A)^{1/2} e_k \rangle L_k = 0 \quad \forall n.$$

Now suppose $\{I, L_k : k \geq 1\}$ are linearly independent. We get $p = 0$, $\langle e_n, a \rangle = 0$ $\forall n$ and $\langle e_n, (I - A)^{1/2} e_k \rangle = 0$ $\forall n, k$. That is, $p = 0, a = 0, I - A = 0$ implying $(A, x, q) = (I, 0, 0)$. Therefore the induced semigroup of τ is same as η. This proves that η is the minimal dilation of τ.

On the other hand if $\{L_k : k \geq 0\}$ are not linearly independent, without loss of generality we can take $\sum_{k \geq 0} c_k L_k = 0$, for some scalars $\{c_k\}$ with $\sum_{k \geq 1} |c_k|^2 = 1$. Let v be the unit vector $\sum_{k \geq 1} \bar{c}_k e_k$. Take $A = I - |v\rangle\langle v|, x = -c_0 v, q = |c_0|^2$. It is easily verified that $(A, x, q) \in \mathcal{D}_{\mathcal{K}}$ (we have $a = -c_0 v, p = 0$). Moreover $\mathcal{L}_G = \mathcal{L}$ as

$$\begin{aligned}
\langle e_n, a \rangle I - \sum_{k \geq 1} \langle e_n, (I - A)^{1/2} e_k \rangle L_k &= -\bar{c}_n c_0 I - \sum_{k \geq 1} \bar{c}_n c_k L_k \\
&= -\bar{c}_n \left(\sum_{k \geq 0} c_k L_k \right) \\
&= 0.
\end{aligned}$$

So the compression to $\mathcal{B}(\mathcal{H}_0)$ is not injective on $\mathcal{D}_\eta$ and therefore η is not the minimal dilation of τ (see remarks made just before Theorem 5.4). ∎

Computations we have made in proving Theorem 9.1 clearly have much more information than a criterion for minimality. To see this we re-write the formula for $\mathcal{L}_G$. Let $\mathcal{K}^+$ be the Hilbert space $\mathbb{C} \oplus \mathcal{K}$, with ortho-normal basis $\{e_0 = 1 \oplus 0, e_1, e_2, \cdots\}$. Let $V : \mathcal{H}_0 \to \mathcal{H}_0 \otimes \mathcal{K}^+$ be the operator defined by, $V u = \sum_{k \geq 0} L_k u \otimes e_k, u \in \mathcal{H}_0$ (where $L_0 = I$). Note that as $\sum_{k \geq 0} L_k^* L_k$ converges strongly V is a bounded operator.

As before fix $(A, x, q) \in \mathcal{D}_{\mathcal{K}}$. Let B be the operator $1_{\mathcal{K}} - [A, X, q]$, that is,

$$B = \begin{bmatrix} q & -x^* \\ -x & (I - A) \end{bmatrix} \quad \text{on } \mathcal{K}^+.$$

Now a routine computation shows that for $G = G(A, x, q)$,

$$\mathcal{L}_G(X) = \mathcal{L}(X) - V^*(X \otimes B)V \qquad X \in \mathcal{B}(\mathcal{H}_0).$$

Take $\mathcal{K}_0 = \overline{\text{span}}\{(X \otimes I)Vu : u \in \mathcal{H}_0, X \in \mathcal{B}(\mathcal{H}_0)\}$. As the projection onto $\mathcal{K}_0$ must commute with $(X \otimes I)$, $\forall X \in \mathcal{B}(\mathcal{H}_0)$, it is of the form $(I \otimes R)$ for some projection R of $\mathcal{K}^+$. Note that $\mathcal{L}_G = \mathcal{L}$ iff $RBR = 0$. Let $\mathcal{K}^-$ be the space $\{e_0, Re_0\}^\perp$ in $\mathcal{K}^+$, and let d be the real number $\langle e_0, Re_0 \rangle$.

Theorem 9.2: Consider the dilation η of τ given by (9.4). Then the associated induced semigroup $\tilde{\tau}$ of τ is given by the projection cocycle $E = G(Q, x, q)$, where

$$Q = 1_{\mathcal{K}^-} R 1_{\mathcal{K}^-}, \quad x = \frac{1}{d} Re_0 - e_0, \quad q = \frac{1}{d} - 1.$$

The deficiency index of $\tilde{\tau}$ is rank of $(1_{\mathcal{K}^+} - R)$, and the index of τ is (rank R)-1.

Proof: To begin with using just the definition of $\mathcal{K}_0$ observe that $d = \langle e_0, Re_0 \rangle$ can't be zero. Now if $0 < d < 1$, let e_1 be the unit vector $(d(1 - d))^{-\frac{1}{2}}(de_0 - Re_0)$. Extend $\{e_0, e_1\}$ to an ortho-normal basis $\{e_0, e_1, e_2, \cdots, \}$ of $\mathcal{K}^+$. It is clear that $(1_{\mathcal{K}^+} - R)$ on that basis is given by

$$1_{\mathcal{K}^+} - R = \begin{bmatrix} (1 - d) & \sqrt{d(1 - d)} & 0 \\ \sqrt{d(1 - d)} & d & 0 \\ 0 & 0 & R_0 \end{bmatrix},$$

where R_0 is a projection in $\mathcal{K}^-$. Then it is not difficult to see that

$$F = \begin{bmatrix} 1/d - 1 & \sqrt{\frac{1}{d} - 1} & 0 \\ \sqrt{\frac{1}{d} - 1} & 1 & 0 \\ 0 & 0 & R_0 \end{bmatrix}$$

is the maximal element of $\{B \geq 0 : 0 \leq 1_{\mathcal{K}} B 1_{\mathcal{K}} \leq 1_{\mathcal{K}}$ and $(1-R)B(1-R) = B\} = \{B \geq 0 : 0 \leq 1_{\mathcal{K}} B 1_{\mathcal{K}} \leq 1_{\mathcal{K}}$ and $RBR = 0\}$. Therefore the projection cocycle for the induced semigroup (we are making use of Corollary 5.5) is given by

$$E = 1_{\mathcal{K}} - F = \begin{bmatrix} -(\frac{1}{d} - 1) & \sqrt{\frac{1}{d} - 1} & 0 \\ \sqrt{\frac{1}{d} - 1} & 0 & 0 \\ 0 & 0 & 1_{\mathcal{K}^-} - R_0 \end{bmatrix} := \begin{bmatrix} -q & -x^* \\ -x & Q \end{bmatrix}$$

In other words, $x = -\sqrt{\frac{1}{d} - 1} e_1 = \frac{1}{d} Re_0 - e_0, q = (\frac{1}{d} - 1) = \|x\|^2$, and $Q = 1_{\mathcal{K}^-} - R_0 = 1_{\mathcal{K}^-} R 1_{\mathcal{K}^-}$. In case $d = 1, Re_0 = e_0$ and therefore $\mathcal{K}^- = \mathcal{K}$, and we have

$$1_{\mathcal{K}^+} - R = \begin{bmatrix} 0 & 0 \\ 0 & R_0 \end{bmatrix}$$

for some projection R_0 of $\mathcal{K}$. This gives maximal element F as $1 - R$, and then

$$E = \begin{bmatrix} 0 & 0 \\ 0 & 1_{\mathcal{K}} - R_0 \end{bmatrix}$$

So that $x = 0, q = 0$ and $Q = 1_{\mathcal{K}} - R_0 = 1_{\mathcal{K}^-} R 1_{\mathcal{K}^-}$. Note that in either case, deficiency index of $\tilde{\tau} = $ rank $(1_{\mathcal{K}} - Q) = $ rank $(1_{\mathcal{K}^+} - R)$, and we have index of $\tau = $ rank $Q = $ rank $R - 1$.   ∎

Corollary 9.3: The set of generators of quantum dynamical semigroups dominated by τ, can be described as

$$\{\mathcal{L}_B : B \in \mathcal{B}(\mathcal{K}^+), B \geq 0, 0 \leq 1_{\mathcal{K}} B 1_{\mathcal{K}} \leq 1_{c\mathcal{K}}\}$$

where $\mathcal{L}_B(X) = \mathcal{L}(X) - V^*(X \otimes B)V$. Moreover there is an action of Heisenberg motion group (of $\mathcal{K}$) on this set given by

$$g_{(U,u,c)}(\mathcal{L}_B) = \mathcal{L}_{M^*BM}, \quad \text{where} \quad M = \begin{bmatrix} 1 & 0 \\ u & U \end{bmatrix}.$$

If $\{L_k : k \geq 0\}$ is a linearly independent sequence, then $\mathcal{L}_B \leq \mathcal{L}_C$ iff $C \leq B$, in particular $\mathcal{L}_B = \mathcal{L}_C$ iff $B = C$.

Proof: Clear from Theorem 9.2, computations of $\mathcal{L}_G$ and remark on the action of Heisenberg motion group in Section 7. ∎

As an immediate corollary of Theorem 9.1 we have the following interesting result of Parthasarathy and Sunder [PSu]. This was a tool in an earlier attempt to prove minimality of dilations [BF]. The proof in [PSu] is through some complicated estimates, and uses Lévy-Doob martingale convergence theorem.

Corollary 9.4 : The closed linear span of exponential vectors

$$\left\{ e(\chi_{[s_1,t_1]} + \cdots + \chi_{[s_n,t_n]}) : 0 \leq s_1 \leq t_1 \leq s_2 \leq t_2 \cdots s_n \leq t_n, n \geq 1 \right\}$$

is $\Gamma(L^2(\mathbb{R}_+))$.

Proof : Take $\mathcal{H}_0 = \mathbb{C}^2$, with standard ortho-normal basis $\{e_1, e_2\}$. Consider the unital quantum dynamical semigroup τ on $\mathcal{B}(\mathcal{H}_0)$ defined by

$$\tau_t\left(\begin{bmatrix} x_{11} & x_{12} \\ x_{21} & x_{22} \end{bmatrix}\right) = \begin{bmatrix} x_{11} & e^{-t/2}x_{12} \\ e^{-t/2}x_{21} & x_{22} \end{bmatrix},$$

$$\text{for } X = \begin{bmatrix} x_{11} & x_{12} \\ x_{21} & x_{22} \end{bmatrix} \in \mathcal{B}(\mathcal{H}_0).$$

Then the generator $\mathcal{L}$ of τ is given by $\mathcal{L}(X) = -\frac{1}{2}[L, [L, X]], X \in \mathcal{B}(\mathcal{H}_0)$ where $L = |e_1\rangle\langle e_1|$. A dilation η of τ can be realized on $\mathcal{H}_0 \otimes \Gamma(L^2(\mathbb{R}_+))$, by taking

$$\eta_t(Z) = U_t^* \tilde{\gamma}_t(Z)U_t, \quad Z \in \mathcal{B}(\mathcal{H}_0 \otimes \Gamma(L^2(\mathbb{R}_+))),$$

where U_t is the Hudson-Parthasarathy cocycle satisfying

$$dU = (LdA^\dagger - L^*dA - \frac{1}{2}L^*Ldt)U.$$

Actually, U_t can be written down explicitly as $U_t = e^{-iL \otimes P(t)}$, where for any real $c, e^{-icP(t)}$ is the Weyl operator $W(c\chi_{t]})$, and U_t can be interpreted using functional calculus (L is self-adjoint).

Then a short computation shows that (see Section 3 of [BF]) for any X in $\mathcal{B}(\mathcal{H}_0)$,

$$\eta_t(X) = \begin{bmatrix} x_{11} & 0 \\ 0 & x_{22} \end{bmatrix} \otimes I_{[0,t]} \otimes \Phi_{[t} + \begin{bmatrix} 0 & x_{12} \\ 0 & 0 \end{bmatrix} \otimes W(-\chi_{t]}) \otimes \Phi_{[t}$$

$$+ \begin{bmatrix} 0 & 0 \\ x_{21} & 0 \end{bmatrix} \otimes W(\chi_{t]}) \otimes \Phi_{[t},$$

where $I_{[0,t]}$ is the identity operator in $\Gamma(L^2([0,t]))$, and $\Phi_{[t}$ is the projection on to vacuum in $\Gamma(L^2([t, \infty)))$. Then by induction we see that for $r_1 \geq r_2 \geq \cdots r_n \geq 0, X_1, X_2, \cdots, X_n \in \mathcal{B}(\mathcal{H}_0), u \in \mathcal{H}_0, \eta_{r_1}(X_1) \cdots \eta_{r_n}(X_n)u$ is in the closed linear span of unions of

$$\left\{ e_1 \otimes e(-\chi_{[s_1,t_1]} - \cdots - \chi_{[s_n,t_n]}) : 0 \leq s_1 \leq t_1 \leq s_2 \leq t_2 \cdots s_n \leq t_n, n \geq 1 \right\}$$

and

$$\left\{e_2 \otimes e(\chi_{[s_1,t_1]} + \cdots + \chi_{[s_n,t_n]}) : 0 \le s_1 \le t_1 \le s_2 \le t_2 \cdots s_n \le t_n, n \ge 1\right\}.$$

Now it is clear that the result follows from the minimality of η.   ∎

Appendix A: Continuity

Here we have some technical results on continuity properties of quantum dynamical semigroups and cocycles of E_0-semigroups. We refer to ([Di], [Su], [Ta]) for definition and basic properties of various topologies of $\mathcal{B}(\mathcal{H})$ used here. Through-out we are abbreviating strong (respectively weak) operator topology to strong (respectively weak) topology.

Propositon A.1: Let $\mathcal{H}_0$ be a complex separable Hilbert space and let $\tau = \{\tau_t : t \geq 0\}$ be a one-parameter semigroup of contractive, completely positive maps on $\mathcal{B}(\mathcal{H}_0)$. Then continuity of all functions $\{t \mapsto \tau_t(X); X \in \mathcal{B}(\mathcal{H}_0)\}$ in the following topologies are equivalent: (i) strong, (ii) σ-strong, (iii) weak, (iv) σ-weak.

Proof : For fixed $X \in \mathcal{B}(\mathcal{H}_0), \{\tau_t(X) : t \geq 0\}$ is bounded in $\mathcal{B}(\mathcal{H}_0)$. On bounded subsets of $\mathcal{B}(\mathcal{H}_0)$ strong (respectively weak) and σ-strong (respectively σ-weak) topologies coincide. Further continuity in strong topology clearly implies continuity in weak topology. So it suffices to show that if the functions listed above are continuous in weak topology then they are continuous in strong topology. We do this as follows: For $u \in \mathcal{H}_0, X \in \mathcal{B}(\mathcal{H}_0), 0 \leq s \leq t < \infty$ taking $Y = \tau_s(X)$, making use of complete positivity we have

$$
\begin{aligned}
\|\tau_t(X)u - \tau_s(X)u\|^2 &= \|\tau_{t-s}(Y)u - Yu\|^2 \\
&= \langle u, \tau_{t-s}(Y^*)\tau_{t-s}(Y)u \rangle - \langle \tau_{t-s}(Y)u, Yu \rangle \\
&\quad - \langle Yu, \tau_{t-s}(Y)u \rangle + \langle Yu, Yu \rangle \\
&\leq \langle u, \tau_{t-s}(Y^*Y)u \rangle - \langle \tau_{t-s}(Y)u, Yu \rangle
\end{aligned}
$$

95

$$- \langle Yu, \tau_{t-s}(Y)u \rangle + \langle Yu, Yu \rangle.$$

Hence $\|\tau_t(X)u - \tau_s(X)u\|^2 \to 0$ as $t \downarrow s$. Similar proof works for t increasing to s. $\blacksquare$

Proposition A.2: Let $\tau = \{\tau_t : t \geq 0\}$ be a semigroup of normal, contractive, completely positive maps of $\mathcal{B}(\mathcal{H}_0)$ for some complex separable Hilbert space $\mathcal{H}_0$, such that functions $\{t \mapsto \tau_t(X) : X \in \mathcal{B}(\mathcal{H}_0)\}$ are continuous in weak topology. Then its minimal dilation $\hat{\tau} = \{\hat{\tau}_t : t \geq 0\}$ is a semigroup of normal, *-endomorphisms of $\mathcal{B}(\mathcal{H})$ for some complex *separable* Hilbert space $\mathcal{H}$ such that $t \mapsto \hat{\tau}_t(Z)$ is continuous weakly (hence strongly) for $Z \in \mathcal{B}(\mathcal{H})$.

Proof: From Proposition A.1, the maps $t \mapsto \tau_t(X)$ are also continuous in various other topologies mentioned there. It follows that $t \mapsto \tau_t(X)Y\tau_t(Z)$ is continuous strongly and hence σ-weakly (we are dealing with a bounded set as $\|\tau_t(X)Y\tau_t(Z)\| \leq \|X\|\|Y\|\|Z\|$) for $X, Y, Z \in \mathcal{B}(\mathcal{H}_0)$. Now from Lemma 3.2 of [AM], $(t, X) \mapsto \tau_t(X)$ is also continuous in product topology (usual topology on $\mathbb{R} \times \sigma$-weak topology on $\mathcal{B}(\mathcal{H}_0)$). (The need of this kind of joint continuity was overlooked in Proposition 5.3 (iii) of [Bh2]. We amend it here).

Now it is clear that functions of the form

$$t \mapsto \langle u, \tau_{s_n}(Y_n{}^* \cdots \tau_{s_2}(Y_2{}^* \tau_{t-s_2}(\tau_{s_1-t}(Y_1{}^*)X\tau_{s_1-t}(Z_1))Z_2) \cdots Z_n)v \rangle$$

$$t \mapsto \langle u, \tau_{s_n}(Y_n{}^* \cdots \tau_{s_2}(Y_2{}^* \tau_{t-s_2}(X)Z_2) \cdots Z_n)v \rangle$$

are continuous in t as t varies in the interval $[s_2, s_1]$ for fixed $s_1 > s_2 > \cdots s_n \geq 0$ and $X_1, Y_1, \cdots, Y_n, Z_1, \cdots Z_n$ in $\mathcal{B}(\mathcal{H}_0)$ for $n \geq 2$. So by Propositions 5.2, 5.4 of [Bh2] we have the result. (These propositions were stated and proved for unital τ, non-unital case requires only minor modifications).

∎

Let $\mathcal{H}$ be a complex, separable Hilbert space. We make $\mathcal{B}(\mathcal{H})$ a measurable space by endowing it with the Borel σ-field coming from its weak (operator) topology. It is same as the Borel σ-field of strong, σ-strong or σ-weak topology. (See comments of Arveson [Ar1], p. 7-8). We recall that algebraic operations, $(X, Y) \mapsto XY, X \mapsto X^*, (X, Y) \mapsto (X + Y)$, multiplication by a scalar etc are measurable. We also note that a function $f : \mathbb{R}_+ \to \mathcal{B}(\mathcal{H})$ is measurable (usual Borel σ-field on $\mathbb{R}_+$) iff the functions $g_{u,v} : \mathbb{R} \to \mathbb{C}$ defined by, $g_{u,v}(t) = \langle u, f(t)v \rangle, u, v \in \mathcal{H}$ are measurable. See Dixmier ([Di], p. 181), for this information. We call such a function f as *weakly measurable* (or simply measurable).

W. Arveson ([Ar1], Proposition 2.5) has shown that if $\theta = \{\theta_t : t \geq 0\}$ is a semigroup of normal, $*$-endomorphisms of $\mathcal{B}(\mathcal{H})$, such that $t \mapsto \theta_t(X)$ is weakly measurable for every $X \in \mathcal{B}(\mathcal{H})$, and $\theta_{t_0} \neq 0$ for atleast one $t_0 > 0$, then $t \mapsto \theta_t(X)$ is weakly (hence strongly) continuous for every $X \in \mathcal{B}(\mathcal{H})$. He has also shown that if a family of isometries $V = \{V_t : t \geq 0\}$ satisfying left cocycle property for θ ($V_0 = I, V_{s+t} = V_s \theta_s(V_t)$ for $s, t \geq 0$) is weakly measurable then it is weakly (hence strongly) continuous. We extend these results to suit our needs.

Lemma A.3: Fix $m > 0$. Let $\Omega = \{A \in \mathcal{B}(\mathcal{H}) : \|A\| \leq M\}$. Suppose $f : \mathbb{R}_+ \times \Omega \to \Omega$ is a function such that for fixed $X \in \mathcal{B}(\mathcal{H}), t \mapsto f(t, X)$ is weakly measurable and for fixed $t, X \mapsto f(t, X)$ is σ-weakly continuous. Then f is jointly weakly measurable, that is, for $u, v \in \mathcal{H}, (t, X) \mapsto$

$\langle u, f(t, X)v \rangle$ is measurable as a function from $I\!\!R_+ \times \Omega$ to $\mathbb{C}$.

Proof: Note that weak, σ-weak topologies on Ω coincide and in weak topology Ω is a compact metric space and in particular separable (see [Di], p. 35). Choose a metric d and a dense set $\{A_k : k \geq 1\}$ for Ω. For $n, k \geq 1$ let $D_{n,k}$ be the set $\{B \in \Omega : d(A_k, B) < \frac{1}{n}\}$. Define $E_{n,k}$, inductively (in k) by $E_{n,1} = D_{n,1}, E_{n,k+1} = D_{n,k} \cap (D_{n,1} \cup \cdots \cup D_{n,k-1})^c$, so as to make the sets disjoint as we vary k for fixed n. Now the functions

$$f_n(t, X) = \sum_k f(t, A_k)\chi_{E_{n,k}}(X)$$

are evidently jointly measurable (χ_E stands for indicator of E). From the continuity of f in the variable X for fixed t, it is clear that $f_n(t, X) \to f(t, X)$, pointwise (weakly), as $n \to \infty$. Therefore f is jointly weakly measurable. ∎

Proposition A.4: Let θ be an E_0-semigroup of $\mathcal{B}(\mathcal{H})$, ($\mathcal{H}$ separable). Suppose $G = \{G_t : t \geq 0\}$ is a family of operators on $\mathcal{H}$ such that (i) $0 \leq G_t \leq 1$ for all $t \geq 0$, (ii) $G_0 = I, G_{s+t} = G_s\theta_s(G_t)$ for $s, t \geq 0$, (iii) G_t commutes with $\theta_t(Z)$ for $t \geq 0, Z \in \mathcal{B}(\mathcal{H})$, (iv) $G_{t_0} \neq 0$ for some $t_0 > 0$, (v) $t \mapsto G_t$ is weakly measurable. Then $t \mapsto G_t$ is strongly continuous.

Proof: Consider the semigroup ψ of normal completely positive maps ψ_t : $\mathcal{B}(\mathcal{H}) \to \mathcal{B}(\mathcal{H})$, Defined by $\psi_t(Z) = G_t\theta_t(Z), Z \in \mathcal{B}(\mathcal{H}), t \geq 0$. Clearly $t \mapsto \psi_t(Z)$ is weakly measurable (strong continuity of $t \mapsto \theta_t(Z)$ has been assumed). Now we can argue with the help of ([HP1], p. 319, Theorem 10.5.5) exactly as done by Arveson in proving ([Ar1], Proposition 2.5). So, it suffices to show that ψ_t is injective for some $t > 0$. Writing θ_t as, $\theta_t(Z) =$

$W_t(Z \otimes 1_{\mathcal{P}_t})W_t{}^*$, as in Section 3, we have $G_t = W_t(1 \otimes \bar{C}_t)W_t{}^*$, for some $\bar{C}_t \in \mathcal{B}(\mathcal{P}_t)$. Taking $t = t_0$, we have $G_{t_0} \neq 0$ and hence $\bar{C}_{t_0} \neq 0$. Then it is clear that ψ_{t_0} is injective. This yields weak continuity of G_t. Strong continuity follows from Proposition A.1 as $G_t = \psi_t(1)$. ∎

Let θ be an E_0-semigroup of $\mathcal{B}(\mathcal{H})$ ($\mathcal{H}$ separable). Suppose that θ is a dilation of a unital quantum dynamical semigroup τ on $\mathcal{B}(\mathcal{H}_0), \mathcal{H}_0 \subseteq \mathcal{H}$. In other words we are having the set up of dilation theory as in Section 3. Note that θ, τ are assumed to be strongly continuous as this is a blanket assumption on all E_0 semigroups and quantum dynamical semigroups in this article. Suppose that $\alpha = \{\alpha_t : t \geq 0\}$ is a semigroup of normal completely positive maps of $\mathcal{B}(\mathcal{H}_0)$ dominated by τ. Assume that for every $X \in \mathcal{B}(\mathcal{H}_0), t \mapsto \alpha_t(X)$ is weakly measurable and there exist $t_0 > 0$, such that $\alpha_{t_0} \neq 0$. Consider $\hat{G}_t, \tilde{G}_t$ constructed as in Section 5.

Proposition A.5: Maps $t \mapsto \hat{G}_t, t \mapsto \tilde{G}_t, t \mapsto \alpha_t(X)$ (for fixed $X \in \mathcal{B}(\mathcal{H}_0)$) are strongly continuous.

Proof: It is clear that maps of the form

$$t \mapsto \langle \hat{\tau}(\underline{r}, \underline{Y})u, \hat{G}_t \hat{\tau}(\underline{r}, \underline{Z})v \rangle$$

are measurable for $(\underline{r}, \underline{Y}, u), (\underline{r}, \underline{Z}, v) \in \mathcal{N}$ in view of (5.1) and Lemma A.3. Then proposition A.4 gives us strong continuity of $\hat{G}_t$, and $\hat{\alpha}$. Now for $x, y \in \mathcal{H}$,

$$\tilde{\alpha}_t(|x\rangle\langle y|) = W_t(|x\rangle\langle y| \otimes \bar{C}_t)W_t^*$$
$$= W_t(|x\rangle\langle a| \otimes 1_{\mathcal{P}_t})W_t^*.W_t(|a\rangle\langle a| \otimes \bar{C}_t)W_t^*$$
$$W_t(|a\rangle\langle y| \otimes 1_{\mathcal{P}_t})W_t^*$$

$$= \theta_t(|x\rangle\langle a|)\hat{\alpha}_t(|a\rangle\langle a|)\theta_t(|a\rangle\langle y|).$$

Hence $\tilde{\alpha}$ is measurable and consequently once again by Proposition A.4, $\tilde{\alpha}, \tilde{G}$ are strongly continuous. The final statement follows as α is a compression of $\hat{\alpha}$. ∎

Finally we have some continuity results applicable for units. We continue with the same set up of τ and θ.

Lemma A.6: Suppose $V = \{V_t : t \geq 0\}$ is a semigroup of bounded operators (with $V_0 = I$) on $\mathcal{H}$, such that $\theta_t(\cdot)$ dominates $V_t(\cdot)V_t^*$, for every t. Suppose $t \mapsto V_t$ is weakly measurable and there exists $t_0 > 0$ such that $V_{t_0} \neq 0$. Then $t \mapsto V_t$ is strongly continuous.

Proof: We have already seen that (see remarks just before (6.1)) $G_t = V_t^* V_t$ is a scalar for every t. Observe that $0 \leq G_t \leq 1$, and $G_{s+t} = G_s G_t$, for $s, t \geq 0$. So $\{G_t\}$ satisfies the local cocycle condition with respect to the trivial semigroup $\gamma_t(Z) \equiv Z$ on $\mathcal{B}(\mathcal{H})$, and Proposition A.4 is applicable. This yields strong continuity of $t \mapsto G_t$. It follows that $G_t = e^{-qt}$, for some $q \geq 0$. Now $W_t = e^{-q/2t}V_t$, is a measurable semigroup of isometries satisfying (left) cocycle condition with respect γ. Then by Arveson's result ([Ar1], Proposition 2.5 (ii)), W (and hence V) is strongly continuous. ∎

Proposition A.7: Suppose $A = \{A_t : t \geq 0\}$ is a semigroup of bounded operators (with $A_0 = I$) on $\mathcal{H}_0$, such that $\tau_t(\cdot)$ dominates $A_t(\cdot)A_t^*$ for every t. Suppose $t \mapsto A_t$ is weakly measurable, and there exists $t_0 > 0$ such that $A_{t_0} \neq 0$. Then $A, \hat{A}, \tilde{A}$ are strongly continuous as maps from $\mathbb{R}_+$ to $\mathcal{B}(\mathcal{H})$. ($\hat{A}, \tilde{A}$ are as in Section 6).

Proof: Weak measurability of $t \mapsto \hat{A}_t^*$ is evident from its definition (6.2). Consequently $t \mapsto \hat{A}_t$ is weakly measurable (note: $\hat{A}_{t_0} \neq 0$ as $A_{t_0} \neq 0$) and from Lemma A.6, it is strongly continuous. Now A is strongly continuous as it is a compression of $\hat{A}$. Finally for $z \in \mathcal{H}$, recalling the definition of $\tilde{A}$,

$$
\begin{aligned}
\tilde{A}_t z = W_t(z \otimes v_t) &= W_t(|z\rangle\langle a| \otimes 1_{\mathcal{P}_t})W_t^* W_t(a \otimes v_t) \\
&= \theta_t(|z\rangle\langle a|)\hat{A}_t a.
\end{aligned}
$$

Hence $t \mapsto \tilde{A}_t$ is strongly continuous.   ∎

Appendix B: Discrete case

Suppose $\tau : \mathcal{B}(\mathcal{H}_0)$ is a normal completely positive map then $\{\tau^n : n \geq 0\}$ (τ^0 being the identity map) is to be considered as a discrete quantum dynamical semigroup. Similarly if $\theta : \mathcal{B}(\mathcal{H}) \to \mathcal{B}(\mathcal{H})$, is a normal, unital, $*$-endomorphism then $\{\theta^n : n \geq 0\}$ is a discrete E_0-semigroup. Consider one such E_0-semigroup. Fix a unit vector $a \in \mathcal{H}$, and take $\mathcal{P}_n$ as range $(\theta^n(|a\rangle\langle a|))$. Then defining unitaries $W_n : \mathcal{H} \otimes \mathcal{P}_n \to \mathcal{H}$, as in the continuous time case (Section 3), we have the decomposition,

$$\theta^n(Z) = W_n(Z \otimes 1_{\mathcal{P}_n})W_n^*, \quad Z \in \mathcal{B}(\mathcal{H}), n \geq 1.$$

Moreover arguments similar to the continuous time yield $\mathcal{P}_n \sim \mathcal{P}^{\otimes^n}$, for $n \geq 1$, where $\mathcal{P} = \mathcal{P}_1$. The dimension of $\mathcal{P}$ is independent of choice of a, and is defined as index of $\{\theta^n : n \geq 0\}$, or that of θ. Now suppose ψ is another $*$-endomorphism of $\mathcal{B}(\mathcal{H})$ such that $\psi(Z) = U\theta(Z)U^*$, for some unitary $U \in \mathcal{B}(\mathcal{H})_0$ then $\psi^n(Z) = U_n\theta^n(Z)(U_n)^*$, where U_n is defined inductively by $U_0 = 1, U_1 = U, U_{n+1} = U\theta(U_n), n \geq 1$. Clearly $\{U_n\}$ is a unitary cocycle of $\{\theta^n : n \geq 0\}$. Now it should be clear that classification of discrete E_0-semigroups up to cocycle conjugacy is a trivial matter as we have just one equivalence class for each index value, $d = 2, 3, \ldots, \infty$, and if $d = 1$ so that we have automorphisms then there is one class each for $\dim \mathcal{H} = 1, 2, 3, \cdots, \infty$. ($\mathcal{H}$ assumed to be separable, non-zero). However, there is some interest in classification of discrete E_0-semigroups up to conjugacy as this problem is intimately connected with classifying representations of Cuntz algebras. For example see Laca [La], Brattelli, Jorgensen, and Price ([BJ], [BJP]), and the references there. Dilation theory has a role to play here as it can be used as a machine to produce interesting examples of $*$-endomorphisms of $\mathcal{B}(\mathcal{H})$. All one has to do is to start with a completely positive map on matrices and dilate. There are also other aspects like possible generalization of intertwining lifting theorem, simultaneous dilation [Bh4] etc.

But generally speaking there isn't much difference between discrete and continuous time as far as dilation theory goes. Concepts like minimal dilation, induced semigroup etc. are there in discrete time as well. Natural discrete analogues of Theorems 2.5, 3.3, 3.6, 3.7, 4.3 etc., hold and can be proved more or less along similar lines.

However naive analogue of factorization theorem fails for the discrete case, namely that not all primary dilations factorize as minimal dilation tensored with some other (discrete) E_0-semigroup. This happens because the index of the tensor product of two E_0-semigroups is now the product of indices unlike the continuous time case where it was the sum. The index of the minimal dilation need not be a factor of the index of primary dilation one starts with. So factorization is not always possible. This being the case, perhaps it is worthwhile to record the correct analogue.

Definition B.1: Suppose $\{\theta^n : n \geq 0\}$ is a discrete E_0-semigroup of $\mathcal{B}(\mathcal{H})$ and is a dilation of a discrete quantum dynamical semigroup $\{\tau^n : n \geq 0\}$ of $\mathcal{B}(\mathcal{H}_0)$. Let $\{\hat{\tau}^n : n \geq 0\}$ be the associated minimal dilation. For any unit vector $a \in \mathcal{H}_0$, take $\mathcal{P} = \text{range } \theta(|a\rangle\langle a|)$, $\hat{\mathcal{P}} = \text{range } \hat{\tau}(|a\rangle\langle a|)$. Then the dimension of $(\hat{\mathcal{P}}^{\perp} \cap \mathcal{P})$ is called as the *deficiency index* of dilation θ of τ (denoted by def (τ, θ)).

We note that in Definition B.1, the deficiency index is independent of the choice of ground vector $a \in \mathcal{H}_0$, and index $(\theta) = $ index $(\hat{\tau}) + $ def (τ, θ). We may define index τ as equal to index $(\hat{\tau})$.

Theorem B.2: Suppose $\{\theta^n : n \geq 0\}, \{(\theta')^n : n \geq 0\}$ are two primary dilations of a discrete unital quantum dynamical semigroup $\{\tau^n : n \geq 0\}$ acting on $\mathcal{B}(\mathcal{H}_0)$. If index $\theta = $ index θ', def $(\tau, \theta) = $ def (τ, θ'), then $\{\theta^n : n \geq 0\}, \{(\theta')^n : n \geq 0\}$ are conjugate as dilation of τ.

Proof: Suppose the two deficiency indices are equal to zero then both θ, θ' are minimal dilations of τ, hence they are conjugate as dilations. So assume that the deficiency indices are non-zero. Now as in the continuous time case dim $(\hat{\mathcal{H}})^{\perp}$ must be infinite dimensional for both θ and θ'. Hence replacing θ' with suitable conjugate dilation we can asume that both θ, θ' are acting on same $\mathcal{B}(\mathcal{H})$, with $\mathcal{H}_0 \subset \hat{\mathcal{H}} \subseteq \mathcal{H}$, in such a way that their compressions to $\hat{\mathcal{H}}$ gives same minimal dilation $\{\hat{\tau}^n : n \geq 0\}$.

Fixing a unit vector $a \in \mathcal{H}_0$, taking $\mathcal{P} = $ range $\theta(|a\rangle\langle a|)$, $\hat{\mathcal{P}} = $ range $\hat{\tau}(|a\rangle\langle a|)$, we have a unitary $W : \mathcal{H} \otimes \mathcal{P} \to \mathcal{H}$ such that W maps $\hat{\mathcal{H}} \otimes \hat{\mathcal{P}}$ unitarily in to $\hat{\mathcal{H}}$, and

$$\theta(Z) = W(Z \otimes 1_{\mathcal{P}})W^*, \quad \tilde{\tau}(Z) = W(Z \otimes 1_{\hat{\mathcal{P}}})W^* \quad Z \in \mathcal{B}(\mathcal{H})$$

$$\hat{\tau}(X) = W(X \otimes 1_{\hat{\mathcal{P}}})W^*, \quad X \in \mathcal{B}(\hat{\mathcal{H}})$$

($\tilde{\tau}$ being the induced semigroup associated with θ). Now as θ' is clearly cocycle conjugate to θ, we have a unitary U on $\mathcal{H}$ such that

$$\theta'(Z) = U\theta(Z)U^* = UW(Z \otimes 1_{\mathcal{P}})W^*U^*, \quad Z \in \mathcal{B}(\mathcal{H}).$$

Further, as $\tilde{\tau}'$ is dominated by θ',

$$\tilde{\tau}'(Z) = UW(Z \otimes E)W^*U^*, \quad Z \in \mathcal{B}(\mathcal{H})$$

for some projection E of $\mathcal{P}$. Take $\hat{Q} =$ range E. Now matching of deficiency indices of θ, θ' (as dilations of τ means that dim $(\hat{Q}) =$ dim $(\hat{P})$, and dim $(\hat{Q}^\perp \cap \mathcal{P}) =$ dim $(\hat{P}^\perp \cap \mathcal{P})$. Hence we can obtain a unitary $V : \mathcal{P} \to \mathcal{P}$ such that $V(\hat{P}) = \hat{Q}$. Let M be the unitary operator $UW(1 \otimes V)W^*$. Then we have

$$\theta'(Z) = MW(Z \otimes 1_{\mathcal{P}})W^*M^*, \quad \tilde{\tau}'(Z) = MW(Z \otimes 1_{\hat{P}})W^*M^*, \quad Z \in \mathcal{B}(\mathcal{H}).$$

Now for $X \in \mathcal{B}(\hat{\mathcal{H}})$ as $\tilde{\tau}'(X) = \hat{\tau}'(X) = \hat{\tau}(X) = \tilde{\tau}(X)$,, we obtain

$$MW(X \otimes 1_{\hat{P}}) = (X \otimes 1_{\hat{P}})W^*MW, \quad \text{for } X \in \mathcal{B}(\hat{H}).$$

$$W^*MW(X \otimes 1_{\hat{P}}) = (X \otimes 1_{\hat{P}})W^*MW, \quad \text{for } X \in \mathcal{B}(\hat{\mathcal{H}}).$$

This means that, restricted to $\hat{\mathcal{H}} \otimes \hat{P}$, W^*MW is of the form $1_{\hat{\mathcal{H}}} \otimes A$, for some unitary $A : \hat{P} \to \hat{P}$. Take $C = MW(1_{\hat{\mathcal{H}}} \otimes (A^* \oplus B))W^*$, where B is any unitary operator on $\hat{P}^\perp \cap \mathcal{P}$. Now clearly $\theta'(Z) = C\theta(Z)C^*$, for $Z \in \mathcal{B}(\mathcal{H})$. Furthermore, for $x \in \hat{\mathcal{H}}, y \in \hat{P}$,

$$
\begin{aligned}
C(W(x \otimes y)) &= MW(1_{\hat{\mathcal{H}}} \otimes (A^* \oplus B))W^*W(x \otimes y) \\
&= MW(x \otimes A^*y) \\
&= W(W^*MW)(x \otimes A^*y) \\
&= W(x \otimes y)
\end{aligned}
$$

As W maps $\hat{\mathcal{H}} \otimes \hat{P}$ unitarily onto $\hat{\mathcal{H}}$, this proves that C fixes every vector of $\hat{\mathcal{H}}$. By similar methods an easy induction argument shows that $\theta^n(C)$ fixes every vector of $\hat{\mathcal{H}}$, for $n \geq 0$. Define $C_n, n \geq 0$, inductively by $C_0 = 1, C_1 = C, C_{n+1} = C_n \theta^{n-1}(C)$. It is clear that $\{C_n : n \geq 0\}$ is a unitary cocycle of $\{\theta^n : n \geq 0\}$, which fixes every vector of $\hat{H}$, and $(\theta')^n(Z) = C_n \theta^n(Z) C_n^*$, for $Z \in \mathcal{B}(\mathcal{H}), n \geq 0$. Now the discrete analogue of Theorem 2.5 proves that θ, θ' are conjugate as dilations of θ.   ∎

We remark that it is not difficult to see that given any unital quantum dynamical semigroup $\{\tau^n : n \geq 0\}$ and any number $d \in \{0, 1, 2, \cdots, \infty\}$ it is possible to obtain a primary dilation of $\{\tau^n : n \geq 0\}$ having d as its deficiency index.

References

[AFL] Accardi, L., Frigerio, A., Lewis, J.T. : Quantum Stochastic Processes, Publ. Res. Inst. Math. Sci, **18** (1982), 97-133.

[AL] Alicki, R., Lendi, K. : *Quantum Dynamical Semigroups and Applications,* Springer Letcure Notes in Phys. **286** (1987), Berlin.

[AM] Accardi, L., Mohari, A. : On the structure of classical and quantum flows, J. Funct. Anal., **135** (1996), 421-455.

[Ao] Aotani, M. : E_0-semigroups in standard form, a product system approach, preprint.

[Ar1] Arveson, W. : *Continuous Analogues of Fock Space*, Mem. Amer. Math. Soc.,**409** (1989).

[Ar2] – : An addition formula for the index of semigroups of endomorphisms of $\mathcal{B}(\mathcal{H})$., Pacific J. Math., **137** (1989), 19-36.

[Ar3] – : Minimal E_0-semigroups, *Operator Algebras and their Applications* (Fillmore, P., and Mingo, J., ed.), Fields Institute communications, AMS, (1997) 1-12.

[Ar4] – : The index of a quantum dynamical semigroup, J. Funct. Anal., **146**, (1997) 557-588.

[Ar5] – : On the index and dilations of completely positive semigroups, Internat. J. Math., **10** (1999), 791-823.

[Ar6] – : Pure E_0-semigroups and absorbing states, Comm. Math. Phys., **187** (1997) no. 1, 19-43.

[Ar7] – : Quantizing the Fredholm index, in *Operator Theory,* Proc. of the 1988 GPOTS-Wabash conference (ed. Conway, J.B. and Morrel, B. B.), Pitman research notes in Math. series, Longman, (1990).

[AW] Araki, H., Woods, E.J.: Complete boolean algebras of type I factors, Publ. Res. Inst. Math. Sci., Series A, Vol **II** (1966) 157-242.

[Be] Belavkin, V. P.: A reconstruction theorem for a quantum random process, (English translation) Theoret. and Math. Phys. **62**, no. 3, (1985) 275-289.

[BF] Bhat, B.V.R., Fagnola F. : On minimality of Evans-Hudson flows, Bull. dell' Unione Mat. Ital., Serie VII, Vol. **IX-A-3p.** (1997) 671-683.

[Bh1] Bhat, B.V.R. : Markov Dilations of Nonconservative Quantum Dynamical Semigroups and a Quantum Boundary Theory, Ph. D. thesis submitted to Indian Statistical Institute on July 6, 1993.

[Bh2] – : An index theory for quantum dynamical semigroups, Trans. of Amer. Math. Soc., **348** (1996) 561-583.

[Bh3] – : Minimal dilations of quantum dynamical semigroups to semigroups of endomorphisms of C^*-algebras, J. Ramanujan Math. Soc., **14** (1999) 109-124.

[Bh4] – : A generalised intertwining lifting theorem, to appear in the Fields Institute Communications volume, *Operator Algebras and their Applications, vol. II*, (ed. Fillmore, P.A., Mingo, J. A.).

[Bi] Biane, Ph. : Calcul stochastique non-commutatif (Non-commutative stochastic calculus), in *Lectures on probability theory* , 1-96, Lecture notes in Math., **1608**, Springer, (1995) Berlin.

[BJ] Bratteli, O., Jorgensen, P.E.T. : Endomorphisms of $\mathcal{B}(\mathcal{H})$ II: Finitely correlated states on $\mathcal{O}_n$, J. Funct. Anal. **145** (1997) 323-373.

[BJP] Bratteli, O., Jorgensen, P.E.T., Price, G. L.: Endomorphisms of $\mathcal{B}(\mathcal{H})$, *Quantization, nonlinear partial differential equations and operator algebra* (Cambridge, MA, 1994), 93-138, Proc. Sympos. Pure Math., **59**, Amer. Math. Soc. (1996), Providence, RI.

[BP1] Bhat, B.V.R., Parthasarathy, K.R. : Markov dilations of nonconservative dynamical semigroups and a quantum boundary theory, Ann. Inst. Henri Poincaré, Probabilités et Statistiques, **31**, no. 4 (1995) 601-651.

[BP2] Bhat, B.V.R., Parthsarathy, K.R. : Kolmogorov's existence theorem for Markov processes in C^*-algebras, Proc. Ind. Acad. Sci. Math. Sci., **103** (1994), 253-262.

[Br] Bradshaw, W. S. : Stochastic cocycles as a characterisation of quantum flows, Bull. Sc. Math. (2) **116**, (1992) 1-34.

[BS] Bhat, B.V.R., Sinha, K.B. : Examples of unbounded generators leading to nonconservative minimal semigroups, Quantum Probability and Applications, **IX**, 89-103, World Scientific, Singapore, (1994).

[CF] Chebotarev, A.M., Fagnola, F. : Sufficient conditions for conservativity of minimal quantum dynamical semigroups, J. Funct. Anal. **153**

(1998), 382-404.

[Ch] Chebotarev, A.M. : The theory of the conservative dynamical semigroups and its applications, Stochastic methods in experimental sciences (Szklarska Porpolhk eba, 1989), 79-95, World Sci. Publishing, (1990) River Edge, NJ.

[CE] Christensen, E. and Evans, D., Cohomology of operator algebras and quantum dynamical semigroups, J. London Math. Soc. **20** (1979), 358-368.

[Da1] Davies, E.B. : *Quantum Theory of Open Systems,* Acad. Press (1976) New York.

[Da2] Davies, E.B. : Quantum dynamical semigroups and the neutron diffusion equation, Rep. Math. Phys. **11** (1977), 169-189.

[Da3] Davies, E.B. : *One-Parameter Semigroups,* Acad. Press (1980), New York.

[Da4] Davies, E.B. : Generators of dynamical semigroups, J. Funct. Anal. **34** (1979), 421-432.

[Di] Dixmier, J. : *Von Neumann Algebras,* North-Holland Publishing Company, (1987) Amsterdam - New York - Oxford.

[EL] Evans, D.E., Lewis, J.T. : *Dilations of Irreducible Evolutions in Algebraic Quantum Theory,* Comm. Dublin Inst. Adv. Stud., Ser. A no.**24** (1977).

[Em] Emch, G. G. : Minimal dilations of CP flows, C^* algebras and applications to physics, Springer Lecture Notes in Math. **650** (1978), 156-159.

[Fa1] Fagnola, F. : Chebotarev's sufficient conditions for conservativity of quantum dynamical semigroups, Quant. Prob. Rel. Topics **VIII**, 123-142, World Scientific (1993) River Edge, NJ.

[Fa2] Fagnola, F. : Unitarity of solutions to quantum stochastic differential equations and conservativity of the associated semigroups, Quant. Prob. Rel. Topics **VII**, 139-148, World Scientific (1992) Singapore.

[FF] Foias C., Frazho A.E., : The commutant lifting approach to interpolation problems, Operator Theory: Advances and Applications, Vol. 44, Birkhauser Verlag (1990).

[GKS] Gorini, V., Kossakowski, A., Sudarshan, E.C.G. : Completely positive dynamical semigroups of n-level systems, J. Math. Phys. **17** (1976), 821-825.

[Gu] Guichardet, A.: *Symmetric Hilbert spaces and related topics,* Springer Lecture Notes in Math. **261** (1972) Berlin.

[HP] Hudson, R.L., Parthasarathy, K.R. : Quantum Ito's formula and stochastic evolutions, Commun. Math. Phys. **93**, (1984) 301-323.

[Jo] Journé, J.L. : Structure des cocyles Markoviens sur léspace de Fock, Probab. Th. Rel. Fields **75**, 291-316 (1987).

[Ku1] Kümmerer, B. : Markov dilations on W^*-algebras, J. Funct. Anal. **63**,(1985) 139-177.

[Ku2] Kümmerer B. : Survey on a theory of non-commutative stationary Markov processes, Quantum Prob. and Appl.-III, Springer Lecture Notes in Math. **1303**, (1987) 154-182.

[La] Laca, M.: Endomorphisms of $\mathcal{B}(\mathcal{H})$ and Cuntz algebras, J. Operator Theory, **30** (1993), 85-108.

[Li] Lindblad, G. : On the generators of quantum dynamical semigroups, Comm. Math. Phys. **48** (1976), 119-130.

[Me] Meyer, P.A. : *Quantum Probability for Probabilists,* Springer Lecture Notes in Math. **1538** (1993), Berlin.

[Mo] Mohari, A. : Quantum Stochastic Calculus with infinite degrees of freedom and its applications (Ph.D. Thesis), Indian Statistical Institue (1991) New Delhi.

[MS] Mohari, A., Sinha, Kalyan B. : Stochastic dilation of minimal quantum dynamical semigroups, Proc. of Ind. Acad. of Sci. (Math. Sci.), **102**, 159-173 (1992).

[Pa] Parthasarathy, K.R. : *An Introduction to Quantum Stochastic Calculus,* Monographs in Math., Birkhäuser Verlag (1991) Basil.

[Po1] Powers, R.T. : An index theory for semigroups of endomorphisms of $\mathcal{B}(\mathcal{H})$ and type II_1 factors, Canad. J. Math., **40** (1988), 86-114.

[Po2] – : A non-spatial continuous semigroup of *-endomorphisms of $\mathcal{B}(\mathcal{H})$, Publ. Res. Inst. Math. Sci. **23** (1987), 1053-1069.

[Po3] – : New examples of continuous spatial semigroups of endomorphisms of $\mathcal{B}(\mathcal{H})$, Internat. J. Math. **10** (1999), 215-288.

[Po4] – : Possible classification of continuous spatial semigroups of *- endo-morphisms of $\mathcal{B}(\mathcal{H})$, Proceedings of Symposia in Pure Math., Amer. Math. Soc., vol. **59** (1996) 161-173.

[Po5] – : Induction of semigroups of endomorphisms of $\mathcal{B}(\mathcal{H})$ from completely positive semigroups of $(n \times n)$ matrix algebras, Internat. J. Math., **10** (1999) 773-790.

[PP] Powers, R.T., Price, G. : Continuous spatial semigroups of *- endo-morphisms of $\mathcal{B}(\mathcal{H})$, Trans. Amer. Math. Soc., **321** (1990), 347-361.

[PR] Powers, R.T., Robinson D. : An index for continuous semigroups of * endomorphisms of $\mathcal{B}(\mathcal{H})$, J. Funct. Anal., **84** (1989), 85-96.

[PSc] Parthasarathy, K.R., Schmidt, K. : *Positive Definite kernels, Continuous Tensor Products and Central Limit Theorems of Probability Theory*, Springer Lecture Notes in Math. **272** (1972) Berlin.

[PSu] Parthasarathy, K.R., Sunder, V.S.: Exponentials of indicator functions are total in the Boson Fock space $\Gamma(L^2[0,1])$, Quantum Probability Communications, **X**, (1998) 281-284.

[Pu] Putnam, C. R.: *Commutation properties of Hilbert space operators and related topics*, Springer-Verlag (1967) New York.

[Sa] Sauvageot, J-L.: Markov quantum semigroups admit covariant Markov C^* dilations, Comm. Math. Phys. **106** (1986), 91-103.

[St] Stinespring, W.F. : Positive functions on C^* algebras,. Proc. Amer. Math Soc. **6** (1955), 211-216.

[Su] Sunder, V.S. : *An Invitation to von Neumann Algebras*, Springer (1987) Berlin.

[SzF] Sz.-Nagy, B., Foias, C. : *Harmonic Analysis of Operators on Hilbert Space*, North-Holland (1970) Amsterdam.

[Ta] Takesaki, M. : *Theory of Operator Algebras* **I**, Springer (1979) New York.

[Th] Thangavelu, S.: *Harmonic Analysis on the Heisenberg Group*, Progress in Mathematics, **159**, Birkhäuser (1998), Boston.

[Vi] Vincent-Smith, G. F. : Dilations of a dissipative quantum dynamical system to a quantum Markov process, Proc. London Math. Soc.(3), **49** (1984), 58-72.

B. V. Rajarama Bhat, Statistics and Mathematics Unit, Indian Statistical Institute, R. V. College Post, Bangalore-560059, INDIA.

e-mail: bhat@isibang.ac.in

homepage: http://www.isibang.ac.in/smubang/bhat

Editorial Information

To be published in the *Memoirs*, a paper must be correct, new, nontrivial, and significant. Further, it must be well written and of interest to a substantial number of mathematicians. Piecemeal results, such as an inconclusive step toward an unproved major theorem or a minor variation on a known result, are in general not acceptable for publication. Papers appearing in *Memoirs* are generally longer than those appearing in *Transactions*, which shares the same editorial committee.

As of September 30, 2000, the backlog for this journal was approximately 11 volumes. This estimate is the result of dividing the number of manuscripts for this journal in the Providence office that have not yet gone to the printer on the above date by the average number of monographs per volume over the previous twelve months, reduced by the number of volumes published in four months (the time necessary for preparing a volume for the printer). (There are 6 volumes per year, each containing at least 4 numbers.)

A Consent to Publish and Copyright Agreement is required before a paper will be published in the *Memoirs*. After a paper is accepted for publication, the Providence office will send a Consent to Publish and Copyright Agreement to all authors of the paper. By submitting a paper to the *Memoirs*, authors certify that the results have not been submitted to nor are they under consideration for publication by another journal, conference proceedings, or similar publication.

Information for Authors

Memoirs are printed from camera copy fully prepared by the author. This means that the finished book will look exactly like the copy submitted.

The paper must contain a *descriptive title* and an *abstract* that summarizes the article in language suitable for workers in the general field (algebra, analysis, etc.). The *descriptive title* should be short, but informative; useless or vague phrases such as "some remarks about" or "concerning" should be avoided. The *abstract* should be at least one complete sentence, and at most 300 words. Included with the footnotes to the paper should be the 2000 *Mathematics Subject Classification* representing the primary and secondary subjects of the article. The classifications are accessible from www.ams.org/msc/. The list of classifications is also available in print starting with the 1999 annual index of *Mathematical Reviews*. The Mathematics Subject Classification footnote may be followed by a list of *key words and phrases* describing the subject matter of the article and taken from it. Journal abbreviations used in bibliographies are listed in the latest *Mathematical Reviews* annual index. The series abbreviations are also accessible from www.ams.org/publications/. To help in preparing and verifying references, the AMS offers MR Lookup, a Reference Tool for Linking, at www.ams.org/mrlookup/. When the manuscript is submitted, authors should supply the editor with electronic addresses if available. These will be printed after the postal address at the end of the article.

Electronically prepared manuscripts. The AMS encourages electronically prepared manuscripts, with a strong preference for $\mathcal{AMS}$-LaTeX. To this end, the Society has prepared $\mathcal{AMS}$-LaTeX author packages for each AMS publication. Author packages include instructions for preparing electronic manuscripts, the *AMS Author Handbook*, samples, and a style file that generates the particular design specifications of that publication series. Though $\mathcal{AMS}$-LaTeX is the highly preferred format of TeX, author packages are also available in $\mathcal{AMS}$-TeX.

Authors may retrieve an author package from e-MATH starting from `www.ams.org/tex/` or via FTP to `ftp.ams.org` (login as `anonymous`, enter username as password, and type `cd pub/author-info`). The *AMS Author Handbook* and the *Instruction Manual* are available in PDF format following the author packages link from `www.ams.org/tex/`. The author package can be obtained free of charge by sending email to `pub@ams.org` (Internet) or from the Publication Division, American Mathematical Society, P.O. Box 6248, Providence, RI 02940-6248. When requesting an author package, please specify $\mathcal{A}\mathcal{M}\mathcal{S}$-LaTeX or $\mathcal{A}\mathcal{M}\mathcal{S}$-TeX, Macintosh or IBM (3.5) format, and the publication in which your paper will appear. Please be sure to include your complete mailing address.

Sending electronic files. After acceptance, the source file(s) should be sent to the Providence office (this includes any TeX source file, any graphics files, and the DVI or PostScript file).

Before sending the source file, be sure you have proofread your paper carefully. The files you send must be the EXACT files used to generate the proof copy that was accepted for publication. For all publications, authors are required to send a printed copy of their paper, which exactly matches the copy approved for publication, along with any graphics that will appear in the paper.

TeX files may be submitted by email, FTP, or on diskette. The DVI file(s) and PostScript files should be submitted only by FTP or on diskette unless they are encoded properly to submit through email. (DVI files are binary and PostScript files tend to be very large.)

Electronically prepared manuscripts can be sent via email to `pub-submit@ams.org` (Internet). The subject line of the message should include the publication code to identify it as a Memoir. TeX source files, DVI files, and PostScript files can be transferred over the Internet by FTP to the Internet node `e-math.ams.org` (130.44.1.100).

Electronic graphics. Comprehensive instructions on preparing graphics are available at `www.ams.org/jourhtml/graphics.html`. A few of the major requirements are given here.

Submit files for graphics as EPS (Encapsulated PostScript) files. This includes graphics originated via a graphics application as well as scanned photographs or other computer-generated images. If this is not possible, TIFF files are acceptable as long as they can be opened in Adobe Photoshop or Illustrator. No matter what method was used to produce the graphic, it is necessary to provide a paper copy to the AMS.

Authors using graphics packages for the creation of electronic art should also avoid the use of any lines thinner than 0.5 points in width. Many graphics packages allow the user to specify a "hairline" for a very thin line. Hairlines often look acceptable when proofed on a typical laser printer. However, when produced on a high-resolution laser imagesetter, hairlines become nearly invisible and will be lost entirely in the final printing process.

Screens should be set to values between 15% and 85%. Screens which fall outside of this range are too light or too dark to print correctly. Variations of screens within a graphic should be no less than 10%.

Inquiries. Any inquiries concerning a paper that has been accepted for publication should be sent directly to the Electronic Prepress Department, American Mathematical Society, P. O. Box 6248, Providence, RI 02940-6248.

Selected Titles in This Series

(Continued from the front of this publication)

For a complete list of titles in this series, visit the
AMS Bookstore at **www.ams.org/bookstore/**.